New World View

New Directions in German-American Studies

Werner Sollors
General Editor

Vol. 7

PETER LANG
New York • Bern • Frankfurt • Berlin
Brussels • Vienna • Oxford • Warsaw

New World View

LETTERS FROM A GERMAN IMMIGRANT FAMILY IN TEXAS (1854–1885)

Edited by Ruth I. Cape

PETER LANG
New York • Bern • Frankfurt • Berlin
Brussels • Vienna • Oxford • Warsaw

Library of Congress Cataloging-in-Publication Data

New World view: letters from a German immigrant family
in Texas (1854–1885) / edited by Ruth I. Cape
pages cm. — (New directions in German-American studies; vol. 7)
Includes correspondence in German, translated into English.
Includes bibliographical references and index.
1. Bergmann, Christian Friedrich, 1817–1888—Correspondence.
2. Bergmann, Christian Friedrich, 1817–1888—Family.
3. German Americans—Texas—Correspondence. 4. Immigrants—Texas—
Correspondence. 5. Farmers—Texas—Correspondence.
6. San Antonio (Tex.)—Biography. 7. San Antonio (Tex.)—
Social life and customs—19th century. 8. Guadalupe River Region (Tex.)—
Biography. 9. Farm life—Texas—Guadalupe River Region—History—
19th century. 10. Guadalupe River Region (Tex.)—Social life and
customs—19th century. I. Cape, Ruth I. II. Title.
F395.G3B45 305.83'107648860922—dc23 2014002277
ISBN 978-1-4331-2652-9 (hardcover)
ISBN 978-1-4539-1328-4 (e-book)
ISSN 1524-7813

Bibliographic information published by **Die Deutsche Nationalbibliothek**.
Die Deutsche Nationalbibliothek lists this publication in the "Deutsche
Nationalbibliografie"; detailed bibliographic data are available
on the Internet at http://dnb.d-nb.de/.

Cover design concept by Corinna Cape

29 Broadway, 18th floor, New York, NY 10006
www.peterlang.com

For

Helga Mittrach

Arlan Bergmann

and

Gene Steele

Contents

Acknowledgments

This edition of correspondence would not have been possible without the dedicated support of Gene Steele and Helga Mittrach, both descendants of the Christian Friedrich Bergmann family. I am grateful to Gene Steele for bringing the Bergmann correspondence to my attention, for providing me with important documents that relate to the letters, and for generously funding the printing of this publication. Furthermore, I am indebted to Helga Mittrach, the owner of the letter collection, for giving me permission to publish the Bergmann letters in this bilingual and annotated edition and for patiently providing useful additional information during our many hours of conversations on the phone.

In the process leading up to publication, students studying German at Austin College in Sherman, Texas gained valuable insight into the letter collection by working on them in two different course projects as part of two advanced German courses. Many students contributed to this edition by diligently entering parts of the German transcription of the letters that were only available as a photocopy into a computer. Thanks go to the following current or former German students of Austin College: Madeline Akers, Christopher Burke, Michael Clark, Helen Hohnholt, Abderrazak Kerfai, Bradley West, Matthew Moore, Randall Ham, Spencer Nystrom, Rebekah Percival, Christopher Stein, Allison Lloveras, Brian Orr, Erika Valek, Sophie Higgs, and Madeline Faye; to Lauren Tepera for turning the floor plan of Bergmann's house into an electronic version. I am also thankful to the German language house assistant for the academic year 2012/2013, Ateeb Rehman, for bringing the original letters from Germany to Texas so they could be consulted to clarify questions that arose during the preparation of the book manuscript. I am grateful to James Kearney for allowing me to consult the manuscript of an article on European Immigration in Texas that, in the meantime, was published in the book *Discovering Texas History* (2014). I am also thankful to one of the editors of the volume, my colleague and former Texas State Historian Light Cummins, for encouraging me to pursue this topic and for readily making James Kearney's article available. Furthermore, thanks go to Bryden Moon for graciously giving me a tour of the former Bergmann land which included a visit to the grave site of C.F. Bergmann; to Frank W. and Ruth Kiel for sharing interesting information about Bergmann's oldest sons Friedrich and Karl in the context of the Civil War, and to the staff of the Patrick Heath Public Library in Boerne, Texas, for access to their archives. I thank

Austin College and the Richardson Foundation for travel grants to visit sites, archives, and libraries in Texas in order to further research the topic of German immigration to Texas and, specifically, the history of the Christian Friedrich Bergmann family. I am also grateful to numerous friends and colleagues on both sides of the Atlantic, especially to Andreas Kiryakakis at St. John's University, MN, for the many inspiring conversations about this collection and his support. Furthermore, I am indebted to Georg Schild at the Eberhard Karls University of Tübingen, Germany, for his encouragement, for reading and commenting on parts of the manuscript, and for sharing his historical expertise about North American history during our many conversations. Finally, I thank Jacqueline Klassen for her diligent editorial work on large portions of the manuscript, and Sharon Jackson for editorial work on the photographs.

Last, but not least, I am indebted to my daughter Corinna Cape for her careful reading of the translation of the letters from German into English, her many useful suggestions and improvements in the English wording, and for designing the book cover concept; to my son Joshua Cape for his support and valuable editorial advice.

Preface

This book is a bilingual and annotated edition of a collection of letters written by the German immigrant Christian Friedrich Bergmann and his family between 1854 and 1885, upon their arrival in Texas. The volume is intended as an insightful and thought-provoking contribution to the larger question of German immigration to Texas in the nineteenth century, and tries to promote a better understanding of today's diverse American culture. At the same time, it is an attempt to show that, historically speaking, 'Americanization' has taken place on both the American and the European continents.

Gene Steele and Helga Mittrach, both descendants of the Christian Friedrich Bergmann family, played an important role in the compilation of this edition. Gene Steele is a great-grandson of Christian Friedrich Bergmann and a grandson of Bergmann's youngest son, Christoph. Helga Mittrach is a great-granddaughter of Christian Friedrich Bergmann's sister. Gene Steele met his cousin Helga Mittrach who resides in Mainz, Germany, for the first time in the fall of 1994 and has been in contact with her ever since. More than twenty years ago, Helga Mittrach, the owner of the original Bergmann letters, decided to transcribe the entire correspondence into modern German font. This endeavor was especially helpful, since fewer and fewer Germans are able to decipher the style of old German script in which the letters were originally written.

The history of these letters is not fully known. For some time, they were in the possession of an organ player in Oberwitz, Germany, who gave them to the sister of Helga Mittrach's mother; from her, Helga's mother received the correspondence in the early 1970s. Eventually, the latter brought the Bergmann correspondence from former East Germany to West Germany. At that time, Helga Mittrach who had left East Germany in 1952, obtained the letters from her mother. However, if it were not for Arlan Bergmann, a great-grandson of Christian Friedrich Bergmann's youngest son Christoph, the content of the Bergmann letters might have never crossed the Atlantic Ocean. Arlan Bergmann who had been eager to learn more about his German ancestors for many years, was not permitted to visit Ebersbach—Bergmann's hometown—while it was part of East Germany. However, shortly after the German reunification in 1990, he seized the opportunity to finally travel to this village. From there he was guided to Helga Mittrach from whom he obtained photocopies of the transcriptions that he later distributed among relatives in Texas.

Several years ago, Gene Steele brought the letter collection to my attention when he took a German course at Austin College from me. Knowing that I was working on another bilingual letter edition at the time, he showed me a photocopy of Helga Mittrach's transcriptions. Eventually, he helped me establish communication with her, and I accepted the invitation to work on an edition of the Bergmann correspondence. As we found out only after I had completed the English translation of all the letters myself, a large portion of fairly good, but at times incorrect or incomplete English translations of Bergmann's letters (translator not listed) had found their way into various issues of *Keys to the Past. A Journal Compiled & Published by the Genealogical Society of Kendall County* in Boerne, Texas. These publications took place without the knowledge of Helga Mittrach. As a result, a comprehensive annotated edition of the correspondence of Christian Friedrich Bergmann and his family had never been authorized or compiled before. Throughout my work on these letters, Gene Steele diligently researched numerous aspects of the family history that he shared with me and that partially have found their way into annotations of this edition.

Today, the land on which Bergmann settled in 1857 upon his arrival in Texas, is part of a gated community, the WaterStone Subdivision near Boerne, Texas. Many descendants of Christian Friedrich Bergmann still live in Texas. As a tribute to Christian Friedrich Bergmann, in 2009, the WaterStone Subdivision renamed their 'North Park' the 'Bergmann Park.' Seventy Bergmann descendants were present at the dedication ceremony.

It was Helga Mittrach's wish that the Bergmann letters would find a home close to where the Bergmann family settled. Therefore, she decided to donate the original letters to the Patrick Heath Public Library in Boerne, Texas.

Ruth I. Cape — Austin College, 2014

Part One:

INTRODUCTION

1. German Immigrants to the United States

Throughout history people have departed from areas where they had lived their entire lives and immigrated to faraway places in search of opportunities to build better lives for themselves and their families. Often dissatisfaction with religious, political, social or economic conditions, political oppression or war at home led them to leave everything behind and take the risk of venturing into largely unfamiliar territory, to begin all over again. In the case of German emigrants, "an adventurous spirit," sometimes referred to as "Wanderlust," might also have played a role in their decision to leave Europe.[1]

One of the destinations of German emigrants was the United States of America. The United States is "a nation of immigrants"[2] and were originally heavily built on immigration from Europe. Many immigrants stayed and succeeded in building new lives, some returned to Europe after they had experienced failure or had reached their financial goal in the New World.[3] In the latter part of the nineteenth century, emigrants who went back to Europe were often referred to as "Americans" in their homelands.[4]

The single largest immigration group in the United States is made up by Germans. They helped shape American society by bringing in their particular farming methods, food, religions, politics, education, and intellectual life. Thus American history has a German element that is "more than a part," but rather "woven into the fabric of America."[5]

2. German Immigrants to Texas in the Nineteenth Century

The nineteenth century was a "century of mass German migration to the New World."[6] In fact, the 1850s were the decade with the highest immigration rate in the entire history of the United States with Germans representing the largest group of immigrants. At the same time, it was also a period when anti-immigrant sentiments peaked.[7] Greater mobility due to industrialization and advancements in transportation, such as steamships and railways that allowed for faster transatlantic journeys and shorter travel times on inland waterways and railways,[8] contributed to large numbers of emigrants from all parts of Germany in the nineteenth century to depart from their homes and settle in America.[9] For some, the destination was Texas. In the 1820s and even more so in the 1830s and 1840s, Texas became the object of "a romantic interest"[10] among German farmers owing to a vast amount of literature that was available in Germany, such as emigrant guidebooks and publications that tried to promote overseas colonization, all praising Texas.[11] There were two major areas where Germans settled: Galveston, Houston and the eastern coast of south Texas and the eastern and central regions of the state, primarily the Hill Country that stretches along the Colorado, Guadalupe, and Brazos rivers.[12]

In Texas persons of German birth or descent are the largest ethnic group of European immigrants. By 1850, a time when many people had left Germany for political and social reasons,[13] Germans represented five percent of the total Texas population, and their number continued to grow throughout the nineteenth century.[14] According to the United States census of 1990, close to three million, or 17.5% of the total Texas population, claimed pure or partial German ancestry.[15]

The first significant wave of German immigration to Texas occurred in the 1830s.[16] At that time, Germans began to settle in what is today known as the "German Belt," an area across the south central part of the state that included the Hill Country, a region in the geographical center of Texas that consists of part or all of twenty-five counties.[17] One of the early German settlers in Texas was the German immigrant Christian Friedrich Ernst Dirks who became known as "Friedrich Ernst." Starting in 1832, Ernst praised the opportunities Texas had to offer in his letters to Germany. Numerous Germans, motivated by the prospect of a better life, decided to leave Europe and settle there.[18]

Between 1836 and 1846, the time of the Republic of Texas, the percentage of Germans in the population increased rapidly, primarily due to immigration, and Germans represented the largest group of European immigrants in Texas, followed by other Central Europeans, among them Czechs.[19] During the 1840s, growing numbers of handbooks were published in Germany, intended to attract more Germans to move to Texas and promising them that they could quickly become citizens, receive free land in Texas, and also were permitted to buy additional land for ridiculously small amounts of money.[20] As a result, the reputation of Texas in the minds of many German farmers became that of a "land of milk and honey."[21]

Much attention has been given to the *Adelsverein*, a group of German noblemen who, in the early 1840s, greatly promoted German immigration to Texas where they wanted to create a German colony. They had come to Texas in search of a solution to the problems of rural overpopulation and political unrest in Germany and created a "chain-migration process" that led many Germans to colonize parts of Texas.[22] More than seven thousand German immigrants arrived in Texas between 1844 and 1847. These were the years when New Braunfels and Fredericksburg, probably the most characteristically German towns in Texas, were founded.[23]

German settlers came from different social, religious, and regional backgrounds: They were peasant farmers, artisans, and intellectuals, Protestants, Catholics, Jews and atheists, Hessians, Prussians, Alsatians, and Saxons.[24] If immigrants survived the sea journey, their futures varied greatly.[25] Some fell victim to epidemics, many settled in urban areas like Galveston, Houston, and San Antonio, and others headed for the Texas Hill Country where they formed the western end of the so-called "German Belt."[26] Around 1850, Texas was home to more than 30,000 German immigrants.[27]

During the 1850s until the Union blockade of Confederate ports during the Civil War, there was another great influx of German newcomers to Texas. By 1880, one third of San Antonio's population was German. Finally, in the 1890s, the influence of German culture in Texas had reached a high, while immigration from Germany to Texas bottomed out and, eventually, stopped after the early 1900s. Only in the aftermath of the Second World War, there was another significant wave of German immigration to Texas.[28]

The two World Wars were quite detrimental to German culture in Texas that had flourished for many years. German-Americans could no longer publicly embrace their bicultural identity. Their national origin had become a contentious issue of national loyalty, and they had to make a final choice: Either they were German or American.[29] Prejudices against Germans diminished interest in their traditions and, in the process, the use of the German language in Texas decreased significantly.[30] By the early 1950s, the German language press that had been thriving for so many years and was crucial to the cultural survival of German heritage in Texas died out.[31]

In the nineteenth century, German culture in Texas flourished. In spite of numerous differences between American and German society, "basic similarities" between the two worlds existed. Both societies were increasingly characterized by "profit-making ventures," and both had "institutions and social arrangements" that, at times, created difficulties for capitalism to succeed.[32] The barrier in Texas was the institution of slavery among Anglo-American colonists who had adopted it although the new Republic of Mexico had forbidden slavery. In Germany it was the continued landholding nobility with special privileges in society, and the institution of the artisan guilds which put restrictions on business pursuits.[33] As Walter Struve observed, both the Old World and the New World "belonged to an expanding world of bourgeois societies" that, at the same time, "interacted with each other in increasing intensity."[34] The latter is also shown in the fact that in the nineteenth century, the two large ports of Galveston, Texas, and Bremen, Germany, developed simultaneously, and their growth was in many ways intertwined.[35]

Germany in the 1850s was still a conglomeration of kingdoms, duchies and principalities that where united as a nation only in 1871 under Otto von Bismarck. Rural emigration from Germany took place primarily for economic reasons, and often whole groups of people from a certain area left together. In some parts, especially in the south of Germany, fragmentation of farms due to inheritance laws made it impossible to make a decent living from farming. In northern and eastern Germany farms were by law handed down to a single heir, thus putting a great burden of finding a place in economic life on those who did not inherit land.[36] Emigration offered them the chance to achieve a better economic future. Since immigrants maintained correspondence with the friends and families they had left behind, their reports about opportunities in Texas sometimes caused others to cross the Atlantic.

In the years before the Civil War, only a quarter of the Texas population did not live on farms. The great majority made a living by depending on agricultural pursuits.[37] Life on the Texas frontier[38] was different from life in Germany in numerous ways. There were fewer roads and fewer towns, and travel distances were great. For the most part, the roads were more or less tracks through woodland and prairies, and wagons and oxcarts could only move slowly. In summer, the soft dirt on the roads made negotiating difficult, and in the fall and spring the vehicles sometimes sank deep into the mud.[39] Settlers lived far apart from each other, so that the next neighbor often lived miles away.[40] As a result, close neighborly relationships as one knew them from the Old World did not exist, and new settlers were prone to being homesick, particularly since communication with family members and old friends back in Germany was quite limited. Besides that, mailing letters was expensive and postal service unreliable. Furthermore, life in the rural areas involved not only hard work, but was also a battle for survival and a constant attempt to keep oneself and one's family out of harm's way in a world where crime and violence were common.[41] The native Indians were a new cultural experience to Germans and frequently posed a threat, especially in areas that bordered Indian Territory.[42] In addition, settlers often encountered unfamiliar and at times frightening animals, for example tarantulas and scorpions, or certain snakes, such as the dangerous water moccasins along the Guadalupe River.[43] They were also surrounded by buffalos, bears, wolves, raccoons, opossums, skunks and wildcats.[44] Finally, an additional challenge for German immigrants was the need to learn English. In general, Germans seemed to have been capable of adapting to their new linguistic environment quite well, even "to the point of endangering the survival of their mother tongue in North America."[45] Nevertheless, there is no doubt that in spite of the existence of large German communities in Texas, settlers, especially in the first years upon their arrival in America, frequently had to overcome language barriers when communicating with Anglo-Americans. German settlers in the coastal areas apparently assimilated to the Anglo-American culture fast, whereas Germans in the Hill Country did so slowly and preserved "German ways," including their language to a degree that non-Germans in the area frequently learned German.[46]

3. State of Research

Comprehensive work on the topic of German immigration to the United States was done by Walter D. Kamphoefner and Wolfgang Helbich. In the early 1980s, Helbich, a North American History scholar at the Ruhr-University Bochum, began to recognize the significance of emigrant letters not only as documents of social history, but also as "a decisive factor triggering emigration,"[47] and initiated the worldwide renowned collection of German immigrant letters, the Bochum Emigrant Letter Collection (*Bochumer Auswandererbrief-Sammlung*, abbreviated BABS) that was microfilmed in 1997 by the Library of Congress. In 2002, the project moved to the Forschungsbibliothek Gotha, Germany, and was renamed to North America Letter Collection (*Nordamerika-Briefsammlung*, abbreviated NABS). The NABS includes all the materials of the original Bochum collection plus the letters that were received later, and it also aims at collecting and making accessible German immigrant letters that were written to the area of the former German Democratic Republic.[48] Today the collection comprises about 10,000 original letters or their copies written by German immigrants between 1820 and 1930.[49] As authentic historical documents about immigration history they are of great scholarly interest. Furthermore, they can also be used as meaningful educational tools in the classroom showing students that Germans, too, used to be immigrants who faced language barriers, were at times discriminated against, and occasionally had conflicts with the native population. Two scholarly letter editions by Helbich and Kamphoefner emanated from this North America Letter Collection, *News from the Land of Freedom. German Immigrants Write Home*[50] and an anthology of immigrant letters written during the American Civil War, entitled *Germans in the Civil War: The Letters They Wrote Home*.[51] It is estimated that from 1820 to 1914, about 250 to 300 million letters crossed the Atlantic from the United States to Germany.[52] The letters in this volume were written by the German immigrant Christian Friedrich Bergmann and his family, who had settled and become farmers in the Texas Hill Country in the second half of the nineteenth century. Their letters are not part of the North America Letter Collection (NABS), but would be a useful addition, particularly since their destination was a village in Saxony, in former East Germany, and letters mailed to that region were not part of the original Bochum Emigrant Letter Collection (BABS).

Regarding the topic of German immigration to Texas, a vast amount of basic information on the subject can be found in the easily accessible digitized "Handbook of Texas Online," a project by the Texas Historical Association.[53] Similarly, the "Portal to Texas History" is a rich online source of readily available materials on German immigration to Texas. It includes collections from Texas archives, museums, libraries, historical and genealogical societies and private family collections. Similarly, the journal *Southwestern Historical Quarterly* frequently addresses the topic of German immigration to Texas. With respect to German immigration to Texas in the nineteenth century, Terry G. Jordan in his book entitled *German Seed in Texas Soil: Immigrant Farmers in Nineteenth-Century Texas* (1966), explored many aspects of German immigrant life in Texas during that time period and described how German immigrants influenced and became influenced by their new environment.[54] His work is a frequent reference in the annotations to the letters in this edition.[55]

Letters of German pioneers in Texas during the nineteenth century have received attention in recent decades, but few comprehensive collections of individual writers have been published. This might partially be due to the fact that in the nineteenth century, letters were written in old German script[56] and, typically, only people of the older generation are still able to read and transcribe these letters into modern German writing. Furthermore, producing complete letter editions is only possible if a corpus of letters has been preserved in its entirety. Oftentimes, only individual letters have survived, or letter collections lie untouched in archives. Of interest to this edition of Christian Friedrich Bergmann's letters is the volume *Die Kettner Briefe. The Kettner Letters*, a bilingual edition of letters written by another early German settler in Texas between 1850 and 1875.[57] Like Bergmann, Franz Kettner was a German immigrant; he arrived in Texas a few years before Bergmann and settled in a similar area, first in Fredericksburg and then further north in Mason County. In contrast to Bergmann, Kettner, who was born in 1826, had immigrated to Texas from the southern German region of Baden in 1848 when he was in his early twenties. His life circumstances were quite different from those of Bergmann who was in his thirties when he crossed the Atlantic and brought along his wife and three sons. As opposed to Bergmann who became and remained a farmer and rancher for the rest of his life, Kettner first tried farming by the Guadalupe River in Comal Country, but shortly after joined a group of Texas Rangers in the Rio Grande country and helped fight Indians and Mexicans. He lived for some time in Gillespie Country where he became sheriff and later was United States assessor in Mason County. Kettner married in 1857 and had three sons and two daughters; he died in 1907.[58] Another rich source of information is Walter Struve's edition of selected letters written by the Gieseke family who had immigrated to Texas in the 1840s.[59] His book explores many important aspects of German immigration to Texas.

4. The C.F. Bergmann Family: Biographical Background

Christian Friedrich Bergmann was a German immigrant who belonged to the large wave of German immigrants to Texas in the 1850s and was part of the "colorful picture of the emigrants of the nineteenth century."[60] By the time he arrived, the "German Belt" in Texas had been well established.

Bergmann was born on April 13, 1817 in Ebersbach, Germany, a small village near Dresden that, at the time, was located in the Kingdom of Saxony.[61] Lutheran Church documents indicate that he married Johanna Christiane Luise Dressler[62] on April 6, 1842, and that they had three sons: Christian Friedrich, Jr.,[63] Karl Heinrich,[64] and Christoph Wilhelm.[65] They began their journey to America in 1854, when Christian Friedrich Bergmann was thirty-seven years old; his wife was thirty-six, their three sons were twelve, nine, and five years old respectively.[66] Notes in their birth certificate records indicate that the Bergmann family emigrated from Germany on August 10, 1854. The family first traveled on the brig *Minna*[67] to Galveston, Texas but could not go ashore due to an outbreak of Yellow Fever. This forced them to continue on a steamer to Indianola, Texas, which they reached on November 6, 1854.[68] Family records indicate they had embarked on their journey in Bremerhaven, Germany, on August 19, 1854.[69] By November 18, the family had arrived in San Antonio, Texas, where they stayed before settling along the Guadalupe River near Boerne, Texas.[70] According to a letter of June 2, 1857,[71] Bergmann bought three hundred and twenty acres of land in Comal County on the Guadalupe River and intended to move to his new land later that year. The area of his property ultimately belonged to the eastern Kendall County and is now part of the WaterStone Subdivision, a gated community.

Bergmann and his family became farmers and ranchers on their land and also were active members of their community.[72] County records indicate that Bergmann became an American citizen in August of 1860. All of Bergmann's sons remained in Texas and lived close to where the family had originally settled. They each married and had their own families. On November 24, 1888, Bergmann died at his residence in Texas at the age of seventy-one.[73] His wife Luise died on January 17, 1902 at the age of eighty-five.[74] Both were buried on their Guadalupe River farm.[75] Bergmann was a German who lived on and improved the land he had bought a few years after his arrival in Texas and, therefore, was a symbol of "locational stability"[76] that was typical for the German farmers in Texas.

Bergmann emigrated at a time when large estates in the northern part of Germany usually were passed on undivided.[77] He came from a well-to-do farm family and had three sisters and at least one brother who took over the family estate in Ebersbach that later, in 1876, was sold.[78] Numerous times in his letters Bergmann requests favors from his father, for example to send him all manner of agricultural equipment, including two wagons, or money. It seems that a lack of funds never really was a major issue in his case. It also suggests that along with information, goods and gear of all sorts could be obtained from the German homeland.[79] Apparently Bergmann's decision to immigrate to America was not caused by despair, but rather by a sense of general dissatisfaction with his life in Germany. Bergmann undoubtedly possessed a sense of adventure and 'Wanderlust,' but he also seemed to have reacted to the limited opportunities Germany offered him when he decided to try to build a better future for himself and his family abroad.[80]

Indeed conditions in Germany in the nineteenth century were challenging. From 1851 to 1854, right before Bergmann immigrated to America, much of Germany was struck by particularly bad potato harvests. Crop failures combined with overpopulation were a big problem in Germany, but by the 1850s, industrialization in Germany also had a negative impact on rural areas since it had expanded to such a degree that factories with mechanical production began to hurt farmers to whom home production was an important source of supplementary income.[81] In the year 1854 when Bergmann set out for America, all of Germany was faced with rising food prices,[82] and the buying power in Germany had hit a low. It is not surprising that 1854 was the year of the "highest migration rate in all of German as well as American history."[83]

5. Importance of the Edition

The collection of letters in this volume is an unabridged bilingual edition of the correspondence that Christian Friedrich Bergmann and his family sent to their relatives in Germany and that reached them. Most letters were addressed to Bergmann's father and other family members. Frequently, individual persons were addressed with regard to specific topics or questions. The first letter in the collection is Bergmann's detailed description of the family's journey to the New World. It includes the trip from their German village Ebersbach to Bremerhaven, the sea journey to Galveston and Indianola, Texas, and also their travels through the prairie to San Antonio upon their arrival in Texas. The next preserved letter to which all three sons added a few lines, is of December 1856. In this letter Bergmann indicates that he had sent three letters to which he had never received a response. The next surviving letter is of June 1857 and includes not only a letter by Bergmann, but also one by his oldest son Friedrich who was almost fifteen years old at the time. Two letters are of 1859, one of January and one of December, and another one of March 1861. There is no preserved correspondence from the time period of the Civil War. The first post-bellum letter is of July 1865, followed by one of June 1866. For 1867 there are two letters, one written in May by Bergmann and one written in the same month by his second-oldest son Karl who was twenty-two years old at the time. In 1871 both Bergmann and his wife wrote letters to a sister-in-law in response to the death of a brother. The collection also includes one long undated letter, most likely written in the early 1870s, in which Bergmann addresses his father, brothers and sisters. In 1879 both Bergmann and his wife wrote letters to a sister-in law, and Bergmann seems to have added two small letters to her two sons in the same mailing. The last two preserved letters were written by Bergmann to his nephews in Germany, one in 1880 and one in 1885. As several comments in the preserved correspondence indicate, Bergmann had sent more letters to his family than they received. Therefore it is safe to assume that some of them never arrived in Germany, probably because they got lost in transit. Similarly, it appears that not all letters that were mailed from Germany to America reached their destination. This letter collection includes the obituary for Christian Friedrich Bergmann, issued by the Germania Farmer Association[84] of which Bergmann had been a member. It gives the reader insight into the way Christian Friedrich Bergmann's life and achievements were perceived by other German immigrants in the area.

In his correspondence Bergmann first describes the family's journey to Texas and then, throughout the course of about thirty years, continues to report about the various stages of his attempts to build a life for himself and his family. His account about the sea journey includes numerous comments about weather conditions, interesting landmarks they passed, and the sighting of unfamiliar birds and fish. Furthermore, he writes about the situation on the ship, such as the kinds and amounts of food they received, fights that occasionally broke out, mishaps, and once how a thief was punished. Often he mentions illnesses from which emigrants sometimes suffered during the journey, deaths of children and adults, and their sea burials. He also remarks that the childless captain of the emigration ship liked Bergmann's five-year-old son Christoph so much that he wanted to buy the child from him. The rest of the Bergmann correspondence is filled with details about a wide range of topics concerning everyday immigrant life in Texas in the nineteenth century. Bergmann describes how, upon their arrival in Texas, he and his wife worked for other people in San Antonio, in order to save up enough money so that they could start farming on their own.[85] Later, he reports expenses associated with his life as a farmer, such as prices for land and animals, but also food, and clothing. In addition, he lists the various plants he grows, and the types, amounts, and prices of crops of his yearly harvests. He also provides in depth information about the characteristics of his land by the Guadalupe River and explains in detail how he built his house. Furthermore, Bergmann comments on the impact of natural catastrophes on his life and ways in which he and his family tried to cope with them. He addresses social problems and crimes like theft and murder, as they occur in the interaction among various social groups in the area: American and German settlers, Indians, and Mexicans. Finally, Bergmann gives occasional insight into the cultural life in Texas.

This edition of Bergmann's letters can at large be grouped with "homogeneous collections" as opposed to "heterogeneous collections;" the latter include materials from many different writers, often on specific topics.[86] A homogeneous collection follows the life of one individual–in this case of one family. In some respects, such a corpus of letters cannot claim to be representative. Nevertheless, the Bergmann letters serve as a powerful authentic voice and allow for valuable conclusions regarding life and customs of early settlers in Texas during the second half of the nineteenth century. They include challenges most settlers faced, for example the widespread crime in many areas. In a letter of 1856, a time when the family still lived in San Antonio, Bergmann writes that he intends to buy land since, in his view, "the most peaceful and best life in Texas is that of a farmer," but he points out that he does not want to live too close to Indian land, even if he had to pay more, due to the tendency of Indians to kill white people and steal horses, mules, oxen, and cows from them. He also comments on other problems in San Antonio, such as the lack of trustworthy people, theft, and widespread alcoholism. The multinational and multicultural atmosphere in San Antonio–he reports that there were even Arabs with camels[87]–seemed to make him feel uneasy at times. Bergmann states that theft and murder in San Antonio were spiraling out of control so that police forces decided to shoot or hang criminals who did not turn themselves in. In a later letter of 1871, written at a time when Bergmann lived on his farm by the Guadalupe River, he mentions horse thefts by Indians again, but points out that he himself had not been a victim of such crimes. He also laments the widespread robbery and murder in Texas in general, as well as attacks on trains.

Since this letter collection spans over a time period of three decades, it presents the reader with important insight into the process of German immigrant acculturation and into challenges many immigrants faced in their attempts of adapting to American culture and making it in the New World. It also contributes to answering the question in which ways German immigrants helped shape American culture. In these respects, the letters do fulfill a somewhat representative function since, inevitably, many German immigrants to rural Texas were subject to similar challenges and responded in comparable ways due to their shared cultural background. In addition, the letters give insight into various aspects of German culture that German settlers tried to preserve in the New World. Among examples for the continuation of German traditions that Bergmann mentions in some of his letters, are the singing societies that were founded by Germans in several towns in Texas in the nineteenth century.[88] As the letters show German settlers often associated with other Germans who had come to the same area, and through their interaction they often carried on with German customs in the New World.

All letters in this collection were personal and addressed to family members in Germany; they were not meant for a larger audience. Through them Bergmann and his family speak to us from within a historical environment and describe observations, achievements, and setbacks of a German immigrant family who came to the New World to start a new life and, as the letters show, also developed a new world view in the context of their acculturation in America. Apart from being personal documents, the letters mirror some of the emotions and thoughts of other German immigrants in Texas. Although early settlers in Texas must have voluntarily accepted the challenges associated with relocating in largely undeveloped territory far away from home and probably had a sense of adventure, there certainly must have be , moments when they felt uprooted, disappointed, and occasionally confused in their new environment. Bergmann does not express any kind of disorientation or uprooting. Yet his lengthy descriptions about how he handled difficult situations indicate that he, too, must have experienced such feelings.

The personal character of the Bergmann letters gives them an almost unrivaled originality because they provide from an immigrant point of view a close and, in many respects, unfiltered look not only at the topic of emigration in general, but also at the historical and social reality of San Antonio and rural Texas in the second half of the nineteenth century in particular. Through the eyes of this newcomer and his family the reader gains insight into a facet of the New World culture from a unique perspective and learns about everyday immigrant life and the many emotional, physical, and intellectual challenges Bergmann and his family faced on a new continent, far away from the place they had left.

Interestingly, we do not hear much about Bergmann's life prior to his emigration since he rarely makes references to it. However, like many other immigrants, Bergmann seemed to conduct an ongoing dialogue between his German past and the new American culture to which he was exposed, in an attempt to navigate through his new life safely and successfully. Thus it does not surprise that he sometimes explicitly contrasts his new surroundings with past experiences in the Old World. These cultural comparisons contribute to a better understanding of the social and cultural climate in Texas during the nineteenth century.

The Bergmann letters are historical documents which readers from various fields might find to be interesting objects for analysis, because they can be examined on different levels and from various points of view. Their pedagogical usefulness in a classroom is hard to deny since incorporating German-American materials, such as immigrant letters not only into history, but also into all levels of language instruction can better meet the needs of an increasingly diverse student body at colleges and universities.[89] Bilingual editions, especially, can serve the purpose of promoting meaningful encounters with the past and possibly aid the revival of the mother tongue of a particular group of immigrants. Thus Bergmann's letters could also be used to educate students by enabling them to examine authentic historical sources both in English and the German original.

6. Limits of Written Communication

Although communication by letter is a means to maintain meaningful relationships with people far away and serves as a bridge between old and new ways of life, it is limited due to the lack of interaction in person. This as an observation the Austrian scholar Gerald Lamprecht made in his monograph entitled *Feldpost und Kriegserlebnis. Briefe als historisch-biographische Quelle* [*Feldpost* and War Experience. Letters as a Historical-Biographical Source, 2001]. Lamprecht developed an insightful methodological framework for studying war letters and examined the way in which a particular soldier who had served in the First and Second World War, processed his war experiences in his letters to relatives in Graz, Austria.[90] Some of Lamprecht's observations can also be applied to immigrant letters since in both instances, there are letter writers who had left behind a familiar world and ventured into fairly unpredictable territory where they at times faced dangerous challenges that they had to overcome in a new and unfamiliar environment.[91] This is also true for Bergmann and his family who lived on the Texas frontier where violence and murder were common. In addition, the Civil War in the early 1860s contributed to further instability in an already challenging life environment.

Written communication cannot completely replace personal conversation during which the language is supported by facial expressions, gestures, tone, and the sound of a person's voice. Thus, interaction that is exclusively based on written correspondence can lead to misunderstandings. In a time of far advanced mass communication technology, one easily underestimates the unsettling impact that prolonged disruptions in communication could cause in former centuries when people lived far apart from each other, especially during historically difficult and dangerous times. It is important to keep in mind the emotional effect a simple letter or a lack thereof might have had on a person on both sides of the Atlantic.

Bergmann mostly describes everyday immigrant life in Texas. His letters are detail-oriented and characterized by frankness and the ability to articulate his experiences well. Although we can learn a great deal about Bergmann from what he expressed in words, we cannot rule out that sometimes he might have created a reality that did not actually exist, but that he produced in the process of writing. At times he might have imposed a type of censorship on himself.[92] Perhaps he described certain challenges as less burdensome or dangerous than they really were because he did not want his family in Germany to worry about him or because he would have considered them a sign of weakness on his part.

Furthermore, Bergmann might have focused in his letters on the positive aspects describing his achievements, acquisitions and challenges he successfully overcame, in order to illicit admiration rather than pity from his family and friends in Germany. It is fairly obvious that he wanted his life in America to be considered a success story. Interestingly, in a letter of 1866, he cautions his brother to immigrate to America, pointing out that this was a decision his brother had to make for himself and for which he would not want to be responsible, and indicating that there were times when he had had second thoughts about having emigrated himself. However, this was a common attitude among immigrants.[93] Emigration inevitably entailed a great risk, and there was no guarantee for success. By cautioning potential German emigrants, immigrants protected themselves from later being blamed that they had given bad advice.

7. Message of the Letters

The Bergmann letters are in many ways a mirror of German immigrant life in nineteenth century rural Texas. They reflect the old and new and give a taste of the multicultural flavor of the New World that was increasingly informed by members who had arrived from the Old World. Many times Bergmann had to learn by trial and error. In his farming efforts he frequently had no other choice than to try out whether certain plants would grow in Texan soil or not and, depending on the positive or negative outcome, he had to adjust his approach to farming. In numerous situations he was forced to be resourceful and use a do-it-yourself approach, for example when he painstakingly made a flail himself so he could thresh wheat. In a letter of 1859, he points out that anyone who had not learned a trade should come to Texas since in Texas he was a shoe maker, wheelwright, cold smith, and a butcher and basically was doing everything that needed to be done. It is obvious that Bergmann felt good about the many new skills he developed and increasingly gained self-confidence through them. This becomes also evident in a letter of December 1859, in which Bergmann explains to his father how he is building his house. It seems like it was a half-timbered (*Fachwerk*) construction that consisted of wood and sturdy stone. As a letter of June 1866 proves, Bergmann was utterly proud of his house which he considered "the most beautiful and best house near and far," and the envy of many. He even hired someone to draw an accurate floor plan that he then sent to his father.

In order to come up with additional sources of income, especially since crops sometimes were bad or failed altogether, Bergmann also had to be creative in developing new skills that would help him make more money. In the early 1870s, he learned the hard way how devastating a disloyal business partner and a natural catastrophe can be after he had gone into business with another German immigrant to make shingles. Bergmann had put up the money and together they had built a shingle machine and cut thousands of shingles that they wanted to sell in San Antonio. However, his business partner took off without paying his share when the shingles were swept away by a great flood of the Guadalupe River.[94]

In his correspondence Bergmann sometimes addresses only his father, other times his father, brothers, sisters and friends, occasionally only a brother, a brother-in-law or sister-in-law, or nephews. It is striking that Bergmann never explicitly addressed his mother in his letters. In fact, the word 'mother' only occurs once, in a letter of December 1856, when he jokingly speculates what his mother might say about the great consumption of butter in his family. This suggests a subordinate role of women in the family. The fact that Bergmann's wife only occasionally added a few lines to her husband's long letters in which she basically confirmed his report, is another sign of the traditional patriarchical structure of the family. The position of patriarch within the family is also expressed in the obituary for Christian Friedrich Bergmann, written by the *Germania Farmer Verein* (Germania Farmer Association), a Texas agricultural association that was founded in 1875 with the purpose to protect cattle from thieves and Indians. Bergmann is remembered by his fellow German settlers for his many virtues, such as his strong character, his hard work, and his achievements.

Making it on the Texas frontier was a challenge and required a great set of human skills, most of all endurance and resilience, adaptability and the willingness to accept risks, but also hard work, resourcefulness, and the ability to recognize good opportunities and to evaluate whether it was necessary to assimilate to the new geographical and social environment. Bergmann possessed many of these qualities. He seems to have been able to make decisions and to follow through on them; he also had convictions. For example, as a supporter of the Union, he did not shy away during the Civil War from hiding soldiers who had deserted on his property.[95] Furthermore, he could be tough on himself. Numerous times, he shows disregard for his own personal needs such as in his later life when his work is more important to him than taking the time to cure an ongoing health problem he had battled for several years. He also appears to have had a good sense of humor that helped him overcome disappointments or setbacks. However, Bergmann's letters do not give deep insight into his emotional life. Their style is mostly matter-of-fact, and he focuses on information about agricultural and social life in nineteenth century Texas. Yet his actions speak for themselves. He is focused on the wellbeing and the security of his family and shows great diligence in giving his three sons the best possible education while considering their different talents and interests.

Interestingly, Bergmann rarely comments on the well-being of his wife and her ability to cope with the challenges of life in the New World. We can only guess how uprooted, lonely, and unsafe she, and probably many female German settlers on the frontier, might have felt at times. Women often spent their days by themselves working hard at home and in the gardens. This was particularly true for less educated women who could only afford a modest life-style. Furthermore, everyday experiences, such as the sounds of birds and insects, especially at night, were foreign to the ears of the German settlers and might have added to a sense of disconnect.[96] In contrast, rural life in Germany was characterized by close ties to family, neighbors and friends, who often lived in close proximity and provided not only regular human address but also a strong sense of belonging. Bergmann's wife Luise mentions in a letter of 1879 that at first she felt quite homesick, but that she no longer would want to trade with her relatives in Germany. Luise, like the wives of many other settlers in the frontier, had to be tough and self-sufficient in many ways.[97] Yet, despite her need to display personal autonomy in many situations in the New World, Luise had to integrate this independence into a family structure that appears to have been patriarchal according to nineteenth century German tradition. Bergmann ruled as head of the household, and all family members followed his guidance and helped together to economically support the family.[98]

Political developments in Germany in the nineteenth century did not play a role among the topics Bergmann addresses in his preserved letters, not even the impact they might have had on the sentiments of German-Americans. In only one letter does he mention a political event in Europe: in a letter of June 1866, he writes that the Civil War atrocities that were committed were probably worse than those that occurred during the French Revolution. In contrast to Bergmann, the German emigrant Frank Kettner, for example, shared his thoughts on the unification of Germany as well as the conservative politics of Otto von Bismarck, the first chancellor of a united Germany after the Treaty of Versailles in 1871. It seems that Kettner read newspapers about political developments in Germany.[99] whereas Bergmann generally does not mention them, perhaps for lack of regular access to newspapers or because as a farmer he was too busy with other tasks. At any rate, Bergmann does not address or allude to the unification of Germany in 1871 that greatly influenced the self-identity of many German-Americans and instilled a sense of pride in them since they suddenly no longer came from just a certain region in Germany, but from Germany as nation.[100]

8. Study Approaches to Immigrant Letters

Bergmann's letters are an important source of information about everyday immigrant life. He provides deep insight into many facets of rural life in the Texas frontier. His letters contain detailed descriptions of expenses associated with his life as a farmer. He writes about the plants he grows, the types and amounts of crops of his yearly harvests, prices for crops, food, clothing, animals, his land, and house construction. He describes natural catastrophes and ways in which he coped with them. Furthermore, he frequently comments on social problems, such as crimes like theft and murder in the area. He also provides information about the structure of relationships between various ethnic groups of people who lived in Texas at the time, the Americans, Indians, Mexicans, and other Germans, as well as social and cultural events, such as dances and song festivals. Other topics are diseases among humans, like outbreaks of Yellow Fever and dysentery, but also diseases among animals. Religion seems to be only of marginal importance to him, although Bergmann mentions in one letter that he was glad to have a Lutheran minister available so his sons could be confirmed.[101] Religious holidays mostly serve as time references.

Studying the psychology in immigrant letters is a rewarding task. Readers might ask how Bergmann and his family were able to survive emotionally in a world that was marked by insecurity and unpredictability.[102] In two letters (December 1856 and January 1859), Bergmann mentions dreams he and others in his family had about Germany; these could be studied as an additional source of information about how Bergmann might have felt. The linguistic aspect of the immigration is another important subject. It is obvious that the Bergmann family did not know much English when they arrived. Thus it could be interesting to examine how or if Bergmann's language in his letters reflects any infiltration by English and how the English he heard around him impacted his style of writing in German in the course of the years. Furthermore, immigrant letters could also be studied as a piece of literature, or even be considered a short epistolary novel. A reader with an interest in creative writing could reflect upon the possible content of the letters to which Bergmann responded.

9. Principles of Transcription and Translation

Upon his arrival in Texas Bergmann and his family regularly wrote letters to family members in Germany.[103] Most letters in this edition were written in black ink, except for a few letters of the 1880s that were written in blue ink. The quality and color of the paper Bergmann used vary. The clear majority of his letters were written on white and only a few letters on light blue stationery. Generally, the letters were dated according to German numerical standards: first the day, then the month, followed by the year. Some letters have the dates listed at the end of the letter. The quality of the handwriting in the letters is, with a few exceptions in some places, consistent and neat.

The grammar in the sentences is sometimes unusual and incomplete in the eyes of the modern reader. This is partially the result of different practices in the nineteenth century pertaining to German grammar and sentence structure, and partially also a lack on Bergmann's part to finish all sentences that he had started in a grammatically logical way. A few times he also seems to have left out words. Furthermore, in some letters he spells English words the way they sounded to him. The 'Guadalupe' River is mostly spelled as 'Guadelupe.' In one letter he spells the word 'Branch' the way a German native speaker who is unfamiliar with English would spell it, namely as 'Braensch.' Occasionally, he even creates new words that are a blend of English and German, for example when he writes about grasshoppers as 'Grashüpper,' instead of 'Grashüpfer.' In order to preserve the flavor of Bergmann's letters and because this does not prevent a proper reading of the German text, such spelling mistakes and grammar flaws were intentionally maintained in the German transcription. However, they are not reflected in the English translation of the letters although they are at times referenced in annotations. Occasionally, Bergmann uses English words and gives their German meanings in parentheses, but later seems to assume that his family will understand them when he writes them.

For the purpose of clarity, a few adjustments in the format of the letters were made. Changes in the punctuation were indicated in many parts of the letters since Bergmann frequently favored endless run-on-sentences. These have been altered. In addition, Bergmann's whimsical or lack of paragraphing has been modified for the purpose of better readability. The letters in this edition are presented in chronological order and, for the sake of clarity, were numbered with the numbers being placed in square parentheses. Equally, occasional minor additions, changes, or explanations by the translator were put in square parentheses. For example, the dates and places of some letters that were listed in the end of the original letters were added to the beginning of the letters in square brackets to enable the reader to better follow the timeline; however, they are still included in the end of those letters.

The German section of this book is a line-for-line transcription of the preserved Bergmann letters, written between the time after his arrival in Texas in November of 1854 and March of 1885. More than two decades ago, a descendant of Bergmann, Helga Mittrach, transcribed the complete letter collection from the old German script, commonly used in the nineteenth century, into modern German script. Hers is altogether an excellent rendition of the original although the original letters had to be consulted numerous times for the purpose of clarification, especially when trying to answer questions with regard to dating the letters or certain words. This edition includes Helga Mittrach's transcription into 'normal German font' with minor corrections and was also used as the basis for the English translation of the letters. In order to preserve the historical character of the text's language, Bergmann's German spelling was at large maintained in the German transcription since it does not prohibit a clear reading and understanding of the text and preserves the personal character of his writing. However, at times nouns that Bergmann wrote in lower case were capitalized in the transcription.

The English translation of the Bergmann letters is a line-for-line translation of the correspondence based on the transcription by Helga Mittrach. The numerical use of numbers in the original letters was generally maintained. However, for reasons of clarity, references to time of day are spelled out, although they were generally given as numerical figures in the original letters. Wherever possible, I tried to present the writers' thoughts in their own words and to use the equivalent of the German that Bergmann and his family employed. Often when Bergmann refers to himself in a group, he mentions himself first. This sequence has been changed in the English translation to make the text more readable. The English translation of the letters includes numerous annotations that were partially taken from secondary literature, partially from interviews with descendants of Bergmann who were willing to share their knowledge and family memories with the translator. All footnotes are intended to provide the English-speaking readership with useful historical, geographical, cultural, and biographical information that is essential for understanding the letters within the context of the time in which they were written. The pages in the index refer to the English section of the edition. Most photographs and visual materials in this book were made available by Helga Mittrach and Gene Steele or are courtesy of the Patrick Heath Public Library in Boerne, Texas.

Notes

[1] Rudolph Leopold Biesele, *The History of the German Settlements in Texas, 1831–1861* (Austin, TX: Eakon Press, 1987), 2. The book is a dissertation of 1928 from the University of Texas at Austin and represents the beginning of the study of German settlements in Texas.
[2] John F. Kennedy, *A Nation of Immigrants* (New York, NY: Harper & Row, 1964).
[3] For more on this topic see the article by Günter Moltmann, "American-German Return Migration in the Nineteenth and Early Twentieth Centuries," in George E. Pozzetta ed., *Emigration & Immigration: The Old World Confronts the New* (New York, NY: Garland Publishing, 1991), 470–84.
[4] Walter Struve, *Germans and Texans. Commerce, Migration and Culture in the Days of the Lone Star Republic* (Austin, TX: University of Texas Press, 1996), 7.
[5] Elliot Shore, "Introduction. A New Look at the Nineteenth Century," in Frank Trommler and Elliot Shore eds., *The German-American Encounter. Conflict and Cooperation between Two Cultures, 1800–2000* (New York and Oxford: Berghahn, 2001), 3–5; 4. For information on famous German immigrants see the *American National Biography* (*ANB*), published by the Oxford University Press, which is also available online; it includes portraits of more than 18,700 men and women, among them numerous Germans who have had a major impact on the shaping of the United States.
[6] Lynne Tatlock, Matt Erlin eds., *German Culture in Nineteenth-Century America: Reception, Adaptation, Transformation. Studies in German Literature Linguistics and Culture Series* (Rochester, NY: Camden House, 2005), xi. The essays in this edition are from a symposium entitled "Transfer Effects: German Culture in Nineteenth-Century Europe," that was held at Washington University in 2004, and looked at 'Americanization' as "a productive resignification, transformation, or re-packing of German ideas, values, and products in the United States," rather than 'Americanization' as "the flow of American ideas, values, money, and products into Europe." (xi)
[7] Walter D. Kamphoefner, Wolfgang Helbich, Ulrike Sommer eds., *News from the Land of Freedom. German Immigrants Write Home* (Ithaca, NY.: Cornell University Press, 1991), 17. In German: *Briefe aus Amerika. Deutsche Auswanderer schreiben aus der Neuen Welt 1830–1930* (München u.a.: Beck 1988). The political movement of "nativism" gained national significance in 1854, when the existing party system collapsed while there was, at the same time, a great influx of immigrants. It was feared, for example, that free rounds of alcohol would buy immigrant votes during political elections, especially among German and Irish immigrants. Frequent prejudices against German-Americans were that they were freethinkers, rationalists and desecrators because they usually enjoyed Sundays with their families and engaged in picnics and festivities that often also involved alcohol. (18)
[8] Walter Struve, *Germans and Texans*, 6–7.
[9] Günter Moltmann, "The Pattern of German Emigration to the United States in the Nineteenth Century," in Frank Trommler and Joseph McVeigh eds., *America and the Germans. An Assessment of a Three-Hundred-Year History*, vol. 1 (Philadelphia, PA: University of Pennsylvania Press, 1985): 14–23; 20–21.
[10] Walter Struve, *Germans and Texans*, 4.
[11] Terry G. Jordan, *German Seed in Texas Soil. Immigrant Farmers in Nineteenth-Century Texas* (Austin, TX: University of Texas Press, 1966), 40.
[12] Walter Struve, *Germans and Texans*, 3.

[13] For many Germans a major setback was the aftermath of the March revolutions of 1848 that had failed to build a more democratic Germany with a constitutional monarchy.
[14] In 1850, the total population of Texas, not including Indians, was 212,295; in 1860, it was 604,215. About a third of the population were slaves. Cf. Randolph B. Campbell, *Gone to Texas. A History of the Lone Star State* (New York and Oxford: Oxford University Press, 2003), 207.
[15] Terry G. Jordan, "*Germans*," Handbook of Texas Online, <http://www.tsha-online.org/handbook/online/articles/png02>, [Accessed May 8, 2013]. Published by the Texas State Historical Association. The easily accessible digitized "Handbook of Texas Online," a project by the Texas Historical Association, provides a vast amount of basic information on the subject of German immigration in Texas. On the topic of the American settlement in Texas, see also Terry G. Jordan, *German Seed in Texas Soil*, 21–30.
[16] Terry G. Jordan, "*Germans*," Handbook of Texas Online, <http://www.tsha-online.org/handbook/online/articles/png02>, [Accessed May 8, 2013]. For the history of German immigration to Texas see also the book by the same author *German Seed on Texas Soil*, 40–59.
[17] Terry G. Jordan, "*Hill Country*," Handbook of Texas Online, <http://www.tsha-online.org/handbook/online/articles/ryh02>, [Accessed May 8, 2013].
[18] Terry G. Jordan, "*Germans*," Handbook of Texas Online, <http://www.tsha-online.org/handbook/online/articles/png02>, [Accessed May 8, 2013]. It is known that an open letter of Friedrich Ernst was printed in North German newspapers. Cf. Walter Struve, *Germans and Texans*, 44.
[19] Walter Struve, *Germans and Texans,* 3.
[20] Walter Struve, *Germans and Texans*, 45. For more information on the details of free land and criteria for acquiring land, see ibid. 45–47.
[21] Terry G. Jordan, *German Seed in Texas Soil,* 40.
[22] Between 1844 and 1846, the *Adelsverein* shipped about 10,000 German emigrants to Texas. Cf. Walter Struve, *Germans and Texans,* 48.
[23] The town of New Braunfels was founded in 1845, the town of Fredericksburg in 1846.
[24] Terry G. Jordan, "Germans," Handbook of Texas Online, <http://www.tshaonline.org/handbook/online/articles/png02>, [Accessed May 8, 2013].
[25] During the long sea journey to America, especially women and small children greatly suffered. The passengers on the ship usually were crowded in the dimly-lit space between decks where fresh air did not circulate and smells easily became unbearable. Cf. W. Phil Hewitt, Terry G. Jordan et alii, *People in Texas* (Boston, MA: American Press, 1981), 7.
[26] Terry G. Jordan, "Germans," Handbook of Texas Online, <http://www.tshaonline.org/handbook/online/articles/png02>, [Accessed May 8, 2013]. The *Adelsverein* was also the founder of the towns of New Braunfels and Fredericksburg. For a map of the western German settlements in Texas in the nineteenth century see Terry G. Jordan, *German Seed in Texas Soil*, 46, fig. 7.
[27] Jürgen Eichhoff, "The German Language in America," in Frank Trommler and Joseph McVeigh eds., *America and the Germans. An Assessment of a Three-Hundred-Year History*, vol. 1 (Philadelphia, PA: University of Pennsylvania Press, 1985): 223–250; 232.
[28] Terry G. Jordan, "Germans," Handbook of Texas Online, <http://www.tshaonline.org/handbook/online/articles/png02>, [Accessed May 8, 2013].
[29] For more on the special situation of Germans in Texas who often formed close-knitted German communities, see John Arkas Hawgood, *The Tragedy of German-America* (New York, NY: Arno Press and The New York Times, 1970), 198–200.
[30] The Germans had been the first group among Texas European immigrants in the nineteenth century to found a newspaper when they created the *New Braunfels Zeitung*. This was followed by a large number of German language newspapers and magazines published in Texas. Cf. W. Phil Hewitt, Terry G. Jordan et alii, *People in Texas*, 50.

[31] Terry G. Jordan, "Germans," Handbook of Texas Online, <http://www.tshaonline.org/hand-book/online/articles/png02>, [Accessed May 8, 2013].
[32] Walter Struve, *Germans and Texans*, 5.
[33] Walter Struve, *Germans and Texans*, 5.
[34] Walter Struve, *Germans and Texans*, 6.
[35] Walter Struve, *Germans and Texans,* 10 and 54–57.
[36] Terry G. Jordan, *German Seed on Texas Soil*, 38. From 1851 to 1854, right before Christian Friedrich Bergmann, the main letter writer of this edition, left Germany, much of Germany was struck by particularly bad potato harvests.
[37] Randolph B. Campbell, *Gone to Texas*, 210–11.
[38] Frontier or "unsettled" territory was defined as "that which had population densities of less than two people per square mile." See "Following the Frontier Line, 1790 to 1890," <https://www.census.gov/dataviz/visualizations/001/> , [Accessed May 8, 2013].
[39] Randolph B. Campbell, *Gone to Texas*, 211. Transportation in Texas was made even more difficult since one could not rely on dependably navigable rivers due to periodic low water levels, floating brush or wood or sand bars. Cf. Randoph B. Campbell, *Gone to Texas*, 213.
[40] Andreas V. Reichstein, *German Pioneers on the American Frontier. The Wagners in Texas and Illinois* (Denton, TX: University of North Texas Press, 2000), 80–81.
[41] See references to crimes and outbreaks of violence in Christian Friedrich Bergmann's letters of 2 June 1857 and 4 January 1879 in this edition.
[42] Early settlers in Texas often experienced that cattle and horses were stolen by Indians, sometimes in an attempt of the latter to drive white settlers from their hunting grounds. Cf. Wayne Gard, "Cattle Rustling," Handbook of Texas Online, <http://www.tshaonline.org/hand-book/online/articles/jbc01>, [Accessed on May 8, 2013].
[43] Theodore Gish ed., *The Diary of Hermann Seele and Seele's Sketches from Texas* (German-Austin, TX: German-Texan Heritage Society, 1995), 456–57.
[44] Theodore Gish ed., *The Diary of Hermann Seele and Seele's Sketches from Texas*, 446–51.
[45] John A. McCarthy, "The German Language in America: An Open Forum," in Frank Trommler and Joseph McVeigh eds., *America and the Germans. An Assessment of a Three-Hundred-Year History*, vol 1 (Philadelphia, PA: University of Pennsylvania Press, 1985), 270–79; 270.
[46] Walter Struve, *Germans and Texans,* 3–4.
[47] Walter D. Kamphoefner, Wolfgang Helbich, Ulrike Sommer eds., *News from the Land of Freedom*, 28.
[48] Wolfgang Helbich, Walter D. Kamphoefner eds., *Germans in the Civil War: The Letters They Wrote Home* (Chapel Hill, NC: The University of North Carolina Press, 2006), xi, n.1.
[49] See the comprehensive and informative database *Auswandererbriefe aus Nordamerika* <http://www.auswandererbriefe.de/neue_sammlung.html>, [Accessed May 8, 2013].
[50] Walter D. Kamphoefner, Wolfgang Helbich, Ulrike Sommer eds., *News from the Land of Freedom. German Immigrants Write Home* (Ithaca, NY.: Cornell University Press, 1991).
[51] Wolfgang Helbich, Walter D. Kamphoefner eds., *Germans in the Civil War: The Letters They Wrote Home* (Chapel Hill, NC: The University of North Carolina Press, 2006; originally published in German with the title *Deutsche im Amerikanischen Bürgerkrieg. Briefe von Front und Farm, 1861–1865* (Paderborn: Ferdinand Schöningh, 2002). Other books on the topic include: Wolfgang J. Helbich: "*Alle Menschen sind dort gleich...*". *Die deutsche Amerika-Auswanderung im 19. und 20. Jahrhundert* (Düsseldorf: Schwann, 1988); Wolfgang Helbich ed., "*Amerika ist ein freies Land ...*" *Auswanderer schreiben nach Deutschland* (Darmstadt: Luchterhand, 1985).

[52] Walter D. Kamphoefner, Wolfgang Helbich eds., *News from the Land of Freedom*, 27 and n. 1. In this book the editors printed twenty letters that they considered "representative of millions of similar letters."

[53] Handbook of Texas Online, <http://www.tshaonline.org/handbook/online>, [Accessed on May 8, 2013].

[54] Terry G. Jordan, *German Seed in Texas Soil* (Austin, TX: University of Texas Press, 1966).

[55] A good source of information on the larger topic of European immigration to Texas in the nineteenth century is James Kearney, "European Immigrant History in the Nineteenth Century," in *Discovering Texas History*, Bruce Glasrud, Light Townsend Cummins, and Cary Wintz eds. (Norman, OK: University of Oklahoma Press, 2014), 94-106. Kearney describes essential characteristics for European immigration. According to Kearney, German immigration to South-Central Texas was for the most part "unorganized," and took place based on the personal decision of individuals and perhaps their families to find the financial means to emigrate and then follow through on their plans.

[56] This old form of German language handwriting is called 'Kurrent' and traces back to late medieval cursive writing. Later, from 1935 until 1941, the so-called 'Sütterlin' script that was used in German schools was based on the 'Kurrent' script.

[57] This edition is an example for a more comprehensive bilingual letter edition. Ilse Wurster and Charles A. Kettner eds., *Die Kettner Briefe: The Kettner Letters: A Firsthand Account of a German Immigrant in the Texas Hill Country 1850–1875* (Wilmington, DE: Comanche Press, 2008). The volume consists of thirty-nine letters exchanged between Franz Kettner and his family in Germany from 1850 to 1875.

[58] Ilse Wurster and Charles A. Kettner eds., *The Kettner Letters*, 3–5. Occasional references to Kettner's observations in this book will give valuable additional insight into the time period during which Bergmann and his family wrote their letters. The Kettner letters are of particular interest as a source of information from within history for the years of the Civil War, since Kettner managed, even during the war years, to get letters by way of Mexico in which he described the situation in Texas to his family in Germany.

[59] Walter Struve, *Germans and Texans. Commerce, Migration and Culture in the Days of the Lone Star Republic* (Austin, TX: University of Texas Press), 1996.

[60] Günter Moltmann, "The Pattern of German Emigration to the United States in the Nineteenth Century," in Frank Trommler and Joseph McVeigh eds., *America and the Germans,* 14–23; 21.

[61] Terry G. Jordan points out, many Texas settlers were Wendish people from Saxon and Prussian Lausitz. Cf. *German Seed in Texas Soil*, 33.

[62] According to birth certificate records, Johanna Christiane Louise Dressler was born on 19 December 1817.

[63] According to birth certificate records, Christian Friedrich, usually referred to as 'Friedrich,' was born on 20 August 1842. He died on 15 June 1919. His grave is in the Boerne Cemetery in Texas.

[64] According to birth certificate records, Karl Heinrich was born on 19 September 1845. A grave stone in the Boerne Cemetery in Texas lists Karl's date of death as 21 November 1938. The first name listed on the gravestone is "Chas." [Charles] H. Bergmann.

[65] According to birth certificate records, Christoph was born on 7 April 1849. He died on 18 May 1917. His grave is in the Boerne Cemetery in Texas.

[66] *Keys to the Past.A Journal Compiled & Published by the Genealogical Society of Kendall County*, vol. XXVIII/1, Spring/Summer 2009, 28–29.

[67] A brig is a sailing vessel that has two square-rigged masts. The *Minna* was built in 1851–52 in the shipyard of St. Magnus near Bremen, Germany. Its port of registry was Blumenthal Flethe. The brig had a cargo carrying capacity of 270 register tons. One register ton is equivalent to 100 cubic feet of cargo space. Cf. Friedrich Spengemann, *Die Schiffe der hannoverschen*

Weserflotte der ehemaligen hannoverschen, später preußischen Ämter Blumenthal, Lehe mit Hafen Geestemünde, Lesum und Dorum und der früheren adelichen Gerichte Lesum und Schönebeck [von der Frühzeit der Seeschiffahrt bis 1901] (Bremen: Vegesack, J.F. Rohr, 1936), 36–37.

68 See the first letter [1.] of the collection. The length of their sea journey was typical for trips from Germany to North America on a slower ship, although already by the 1840s, sailing on a superior vessel the Atlantic could be crossed in fourteen days. The majority of emigrants traveled on the slower vessels. Cf. Walter Struve, *Germans and Texans*, 7.

69 *Keys to the Past*, vol. XXVIII/1, Spring/Summer 2009, 29.

70 On the history of Boerne in this time period, see Jefferson Morgenthaler, *The German Settlement of the Texas Hill Country* (Boerne, TX: Mockingbird Books, 2007), 107–125. Although the book is rich in information, unfortunately, it does not contain any footnotes, but just a general bibliography.

71 See letter [3.1].

72 Christian Friedrich was a Grand Juror in 1863, 1868 and 1869. In 1867, he served in the Road #9 Crew. In 1870, he became County Commissioner. The names of Bergmann, his oldest son Christian Friedrich, Jr. and his youngest son Christoph are listed among the original thirty-six charter members of the *Germania Farmer Verein* at Anhalt of which Bergmann was its first elected Vice-President. See *Keys to the Past*, vol. XXVII/1, Spring/Summer 2008, 23.

73 His date of death is also listed on the Indianola Immigrant database, <http://vrhc.uhv.edu/manuscripts/indianola/iddresult.cfm?keyword=Christian+Friedrich+Bergmann&submit=Search>. [Accessed May 11, 2013].

74 *Keys to the Past*, vol. XXVII/1, Spring/Summer 2008, 25.

75 Their graves are located in what is today the WaterStone Additon in Boerne, TX

76 Terry G. Jordan, *German Seed in Texas Soil,* 193.

77 Walter D. Kamphoefner, Wolfgang Helbich, Ulrike Sommer eds., *News from the Land of Freedom*, 3; another important role for emigration at the time were political reasons.

78 A handwritten note on the original wooden frame with a photograph of the farm states: "Ostern 1876, verkauft für 24.000 Mark. Grundbesitz: 44 Scheffel." (Easter 1876, sold for 24,000 Mark, property of 44 bushel.) A 'Scheffel' was an old German measure of capacity and the equivalent of land needed to plant one 'Scheffel' of seed. A 'Dresdner Scheffel' was about 103.8 L or 3 bushels. Depending on the region, one 'Scheffel' was about 12 to 42 ares (1 are equals 100 square meters).

79 This has also been confirmed with respect to other German settlers. Cf. Walter Struve, *Germans and Texans*, 9.

80 According to Helga Mittrach, Bergmann's youngest son Christoph had told his grandfather when they departed that they would soon be back.

81 Terry G. Jordan, *German Seed in Texas Soil,* 38. Apparently, the Bergmann family also had a loom in their house. See also Theodore Huebener, *The Germans in America* (Philadalphia and New York: Chilton Company, 1962), 134–35.

82 Walter D. Kamphoefner, Wolfgang Helbich, Ulrike Sommer eds., *News from the Land of Freedom*, 149.

83 Walter D. Kamphoefner, Wolfgang Helbich, Ulrike Sommer eds., *News from the Land of Freedom*, 10. After 1857 the economic situation in the United States worsened which led to a lower rate of immigration.

84 The *Germania Farmer Verein* (Germania Farmer Association) was an association founded in 1875 with the purpose to protect cattle from thieves and Indians. The cattle were branded by the residents with the association brand G and their efforts were successful. The association received a state charter by 7 February 1876. Cf. David DeKunder, "Anhalt Hall," Handbook of

Texas Online, <http://www.tshaonline.org/handbook/online/articles/xda04>, [Accessed May 8, 2013]. See also n. 72 above.

[85] It was not uncommon that newly arrived settlers worked for Germans who had arrived in Texas earlier, in order to earn money and to gain experience. Cf. Terry G.Jordan, *German Seed on Texas Soil*, 186.

[86] In German, "homogene Sammlungen," as opposed to "heterogene Sammlungen." Cf. Martin Humburg, "Deutsche Feldpostbriefe im Zweiten Weltkrieg," in *Andere Helme—Andere Menschen? Heimaterfahrung und Frontalltag im Zweiten Weltkrieg. Ein internationaler Vergleich*, Detlef Vogel and Wolfram Wette eds., (Essen: Klartext, 1995), 19.

[87] In April of 1856, thirty-five camels were imported from Africa and landed in Indianola, from where they were taken to Camp Verde, northwest of San Antonio. Within a year, another forty-one camels arrived. One considered them suitable beasts of burden on the Texas frontier, since they needed little water and could carry up to six hundred pounds over long distances. In the time before the Civil War, the army successfully used them for exploring the Big Bend area and on expeditions to New Mexico, Arizona, and California. Cf. Randolph B. Campbell, *Gone to Texas*, 199.

[88] The singing society of San Antonio was formed in 1852, the singing societies of Fredericksburg and Comfort were founded in 1858. Cf. R. L. Biesele, *German Settlements in Texas,* 223. The so-called "Sängerfeste," festivals of the singing societies, were held until the Civil War started and revived when the war was over. Cf. Gilbert Giddings Benjamin, *The Germans in Texas. A study in immigration* (Austin, TX: Jenkins Publishing Company 1974; originally published 1909), 118.

[89] On this topic see Gisela Hoecherl-Alden and Maedelon Köhler-Busch, "Expanding on Authenticity: A Case for Immigrant Letters, Photographs, and Other German-Americana in the Language Classroom," *Unterrichtspraxis/Teaching German*, vol. 31/1 (1998), 38–46.

[90] Gerald Lamprecht, *Feldpost und Kriegserlebnis. Briefe als historisch-biographische Quelle* (Innsbruck: Studien Verlag, 2001).

[91] See also Ruth I. Cape, *Youth at War. Feldpost Letters of a German Boy to his Parents, 1943–1945* (New York, NY: Peter Lang, 2010), 3–25.

[92] Lamprecht calls it an "internal censorship." See Gerald Lamprecht, *Feldpost und Kriegserlebnis,* 43.

[93] See more on this topic in Walter D. Kamphoefner, Wolfgang Helbich, Ulrike Sommer eds., *News from the Land of Freedom,* 29.

[94] See letter [11.1].

[95] Interestingly, Bergmann's oldest son Friedrich "claimed Confederate army service in the 1910 U.S. Census. Equally, Karl's obituary states that "during the Civil War he served for and held faith with the South." See Frank Wilson Kiel, *Civil War Soldiers in Kendall County–A Biographical Dictionary* (Comfort, TX: Skyline Ranch Press, 2013), 29. Both Friedrich and Karl seemed to have been supporters of the Confederacy. It is safe to assume that there were some dissentions in the Bergmann household with regard to the Civil War. In their book *Germans in the Civil War*, Helbich and Kamphoefner present three hundred and forty-three letters written during the Civil War by seventy-eight German immigrants of fifty-seven different family groups. According to Helbich and Kamphoefner, about 200,000 soldiers or about ten percent of the Union army, were Germans. Cf. *Germans in the Civil War,* xi. With their edition Helbich and Kampfhoefner showed that immigrants "brought different values, images, and experiences to bear on the war, which provide a contrast to the perspectives shared by most American letter-writers." See *Germans in the Civil War,* xii.

[96] As Reichstein points out this made it especially hard on women to feel at home because they spent the majority of time at home while the men were more distracted with their daily work. Cf. Andreas V. Reichstein, *German Pioneers on the American Frontier*, 80–81.

[97] Despite the fact that German women in rural areas, traditionally, were subservient to men, they played many important roles in the domestic sphere: they carried out work in or near the home and were largely in charge of preparing and storing food. Usually, they were also responsible for spinning the thread and weaving the fabric from which they made the clothes for their families. As mothers they were the main caregivers for the children and essential in maintaining German traditions. Cf. Linda Schelbitzki Pickle, *Contented Among Strangers: Rural German-Speaking Women and Their Families in the Nineteenth-Century Midwest* (Urbana and Chicago, IL: University of Illinois, 1996), 25.
[98] The text of the obituary in this edition also reflects this family structure. Although women were expected to be strong and hardworking, they also were supposed to be submissive. Antebellum Texas was said to be "heaven for men and dogs and hell for women and oxen." See Randolph B. Campbell, *Gone to Texas*, 231. Legally, women had significant property rights, such as community property rights to things that were acquired while being married and also the right to the full title to property after the deaths of their husbands. Cf. Randolph B. Campbell, *Gone to Texas*, 231.
[99] Kettner mentions in a letter of 15 March 1858 that in Fredericksburg he has access to news in Europe since they receive two newspapers from New York and one Texas newspaper. Cf. Ilse Wurster and Charles A. Kettner eds., *The Kettner Letters*, 64. In 1860 there were three dailies, three tri-weeklies, and sixty-five weekly papers in Texas. Most of them had only four pages, but they published local, national and international news. Cf. Randolph B. Campbell, *Gone to Texas*, 230.
[100] For more information on this topic see Hans L. Trefousse, "The German-American Immigrants and the Newly Founded Reich," in Frank Trommler and Joseph McVeigh eds., *America and the Germans. An Assessment of a Three-Hundred-Year History*, vol 1 (Philadelphia, PA: University of Pennsylvania Press, 1985), 160–75.
[101] See letter [3.1] of 2 June 1857.
[102] In this context see the thought-provoking observations on psychoanalytical approaches to explain emigration presented by Andreas V. Reichstein, *German Pioneers on the American Frontier*, 208–210.
[103] In more recent years, without the knowledge of the owner of the original of the Bergmann letter collection a fairly large portion of Bergmann's letters was published in English in four different issues of the local publication *Keys to the Past. A Journal Compiled & Published by the Genealogical Society of Kendall County* (author of translation not listed). These translations can be found in the following issues: vol. XXIV/2 Fall/Winter 2005; vol XXV/1 Spring/Summer 2006; vol. XXV/2 Fall/Winter 2006; vol. XXVI/1 Spring Summer 2007; vol XXVIII/2 Spring/Summer 2009. The published sections are fairly good, yet in some places inaccurate and incomplete renditions of the German original.

Part Two:

THE BERGMANN LETTERS

1. The Sea Journey to America: August to November 1854

[1.]
[Description of the sea journey to America in 1854; undated letter, probably late 1854/beginning 1855][1]

Dear Father:
Trembling and haltingly I take my quill, wondering whether my letter will find you alive or not. I am sure you longed to receive a letter from me, because I promised to write you as soon as I would stand on solid ground again. But, unfortunately, sending a letter costs a lot and since I knew Heuchling[2] had written, I hoped you would find out that my family and I have reached the shore safe and sound.

Now I would like to tell you about my difficult and dangerous journey. On the first day after my farewell, we came to Leipzig[3] and spent the night in the *Schwarze Kreutz*.[4] Dear brother-in-law, you know how it is in Leipzig and everywhere all the way to Bremerhaven,[5] especially regarding emigrants because they have money.[6] On the second day, we stayed in Braunschweig,[7] where they had a fair at the time. On the third day at four o'clock in the afternoon, we arrived in Bremen.[8] We had to stay there until the sixteenth [of August]. Then we took the steamboat *Roland*[9] to Bremerhaven where we had to go to the Immigration House.[10] On our way there we were stranded around Vegesack[11] because we had almost 500 passengers aboard, and were stuck for 3 hours. Since there was a high tide, we only moved on close to seven o'clock. Finally, at eleven o'clock in the evening, we arrived in Bremerhaven, where we were welcomed by some fairly hard beds.

On the 19th our ship, *Minna Blumenthal,*[12] was towed out of the harbor by the steamboat *Magnat*.[13] Because of the annual fair, we lay at anchor until the 21th. On the 21th towards ten o'clock, the anchors were lifted, and we set out to sea in good wind and with full sails. It was interesting how the Weser River and the Baltic Sea [North Sea][14] were fighting with each other in a way that turned the two elements into a shiny blend, so that we saw 3 types of water. On the 22th the wind became stronger and, on the 25th almost developed into a storm. Towards the evening, it became so powerful that the boxes in the tween deck tumbled upon each other and Seiler's chest fell on Christoph[15] so that we thought both of his legs were shattered. Karl[16] was also knocked about, and both of them had upset stomachs since all three of us had become seasick due to the surge. Towards midnight things calmed down, and on the 26th–27th [of August], there was no wind at all.

On the 28th I went on deck before sunrise—a beautiful morning. The sun was shining so brightly, and we were in the Channel. The chalk cliffs towered on the English coast like the cliffs of Saxon Switzerland.[17] There was no wind from eight o'clock in the morning until three o'clock in the afternoon. Towards evening, an English warship[18] showed up, and I counted another 61 vessels in the Channel. Around midnight, I saw to the left of the deck an English city that lay most beautifully illuminated right in front of my eyes. Early on the 29th [of August], a small bird, like a siskin, climbed up the hemp rope and ate some insects. In the morning and in the evening, we also heard a roar of cannons from the French coast.

Early on the 30th we saw an English screw steamer[19] that kept sailing around in the Channel. As our captain told us, it was an observation ship. We also saw many large fish, called *Stier*.[20] Towards evening, we saw a spider crab that looked like a skull when it emerged from the water. There was also a school of small fish. We had little wind in the Channel. Only around noon of the 31th, we caught morning wind, so that all sails were hoisted and our *Minna* carried us out to sea faster. On September 1, we still had good wind and towards eight o'clock in the morning, we again saw the English coast, which consisted of mountains that were partially wooded, partially arable. We also saw cliffs with lighthouses, towns, and villages on them. At sunset, we encountered a Swedish ship with four masts and left the Channel. On the 2nd we had good wind and saw two kinds of fish: pigfish[21] and *Nordkapp.*[22] The sailors said that the latter were too large to catch with harpoons.

Early on the 3rd, two sharks showed up accompanied by small fish (dolphins). Towards one o'clock, a French brig approached from the west. It came from Africa. A barge with four of its sailors came to our ship in order to get food supplies since theirs had been lost in a storm. Tonight a woman died who had already been sick when she boarded the ship. She was sewn into a sack of straw. The sailors did that, and the next evening, she was committed to the surging waters while prayers were said. On the 4th we passed a brig, and a petrel bird that looked like a ringdove sat on the hemp rope. On the 5th we saw two big pigfish. From the 1st to the 5th [of September], there was south eastern wind. On the 6th the wind turned to the west and was also favorable for us.

From the 3rd–8th [of September], my wife was quite ill. She had terrible headaches and pain in her lower abdomen so that we thought she might have dysentery.[23] On the 7th the wind was not in our favor. On the 8th I became very sick. I had attacks of fever, terrible headaches and pain in my lower abdomen. Only on the 13th did I begin eating again. I did not write while I was sick, and there was nothing going on anyway, except for disputes and even worse, fist fights. On the 17th a newborn baby died. That evening we passed the Canary Islands,[24] but we did not see them. On the 18th another child died of a fever of dentition;[25] it was 5 months old, and its body was committed to the sea that evening.

From then on, we had good wind until the 23th. We gained 10 German miles[26] in four hours. During that time, we saw different fish: flying fish, dolphins, etc. From the 23th–26th there was no wind. On the same day we saw a brig to our left. On the 27th two ships were approaching, a brig, and a barge. We saw big fish, called *Speerhai*.[27] They all were 5 to 6 cubits long. On the evening of the 29th, we saw a large bird, similar to a heron,[28] and on the 30th a big black bird showed up that looked like a raven.[29] On the 1st of October, we still had good wind. In the evening, we had our second dance.

Starting on the 18th [of September], there was very little wind, and it was quite hot. I have to mention that up to that time we had experienced very little rain. Whenever it rained on a day we also got sunshine that same day. The weather out at sea is completely different from that ashore. We sometimes wished it would rain to get drinking water, of which we had very little. For 5 people we hardly had 3 pints, and the water was awful! I had to think of the well by the pond and the backmost meadow, but, unfortunately, we were not there. The position of the sun close to Cuba[30] is almost vertical. That is the reason why it is so hot and the water in the barrels gets warm and smelly. Yet, our 3 boys liked it better than the best spring water back home. Christoph had been doing rather well in recent times. He had become popular with the sailors, also with the cook, even the captain wanted to have him. The latter wanted to buy him because he did not have a child of his own.

By the way, I have to mention that after such a long journey, we all were yearning for the mainland, like a fish that has been carried in warm weather for hours craves water. On the 5th [of October], they caught a thief who had stolen a few small items. The captain grabbed him by his arm while the boatswain (petty officer) whipped his behind 10 times with a hemp rope. In the evening, we saw thunderstorms on both sides, but the rain only reached us early on the 6th, so that our water barrels were filled with rain water. We, too, filled all containers we had. That helped against thirst for a few days. On that *day, we saw a beautiful butterfly. On the 7th we had no wind and extreme heat;* we had 28 degrees [Celsius][31] in the shade and 38 degrees [Celsius][32] in the sun. On the 10th there were thunderstorms until midnight. On the same day, we saw different species of beautiful water fowl. On the 11th a swallow, similar to your barn swallows, arrived and accompanied us for several days. On the evening of the 12th, we had thunderstorms and strong winds, and more thunder on the 13th.

On the 14th [of October], we saw two islands, Antigua[33] and Montserrat.[34] The latter was a wonderful sight, for we passed it very closely. Down at the beach was a little town and on the mountains there were tobacco plants. We also saw tube palm trees through the big flue, and here and there houses, and all passengers were delighted by this. In the evening, we were surrounded by thunderstorms and gusty winds. Therefore, all the sails were reefed, except for three because thunderstorms were common there, I mean really strong thunderstorms. After all, we were only 16–17 degrees away from the equator. Our captain made a detour of 100 German miles.

Early on the 15th [of October], a child was stillborn, and it rained for the whole day. On the 16th a two-year-old child died. It had been sickly during the whole trip. We also caught a flying fish. It had an almost rectangular shape. On the 16th due to unfavorable winds, we saw Puerto Rico.[35] On the 17th several flocks of migratory birds came by. They were so hungry that while I ate lunch with my children, one of them sat on my head, the other one on the rim of the bowl. They stayed aboard for several days.

On the 18th a six-year-old boy died of nervous fever[36] and was committed to the foaming waves in the evening. On the 20th a water fowl the size of a duck was shot when it sat on the prow. Before noon, we saw Domingo[37] with a quite visible mountain rising 6,000 feet above sea level, and only on the 24th did our eyes lose sight of it. On the 21th a two-year-old girl died of a fever of dentition.[38] Again, we saw large fish, called *Ungeheuer*.[39] At nine o'clock in the morning of the 23rd, we passed a ship that had suffered a wreck several days earlier during the hurricane near Galveston.[40] The captain said that such things occur in this region. He told us it was a ship with three masts and that this was the reason why he always had the sails lowered during thunderstorms. I am sure you can imagine how frightened we all were when we saw this.

Around noon on the 24th [of October], we saw Jamaica,[41] which has a length of 40 geographical miles[42] and presented itself to our eyes in a large range of high mountains.[43] We sailed by them in the afternoon of the 25th. On the 27th another child died of a fever of dentition and swelling of the gums.[44] It was one and a half years old. That same morning we passed the south end of Cuba, called Kobantan, where a colossal lighthouse towered that we passed already at midday.[45] Then we arrived in the Gulf of Mexico. On the 28th

Heuchling's youngest girl died. From the 29th–30th, there was no wind. On the 31st of October and on the 1st–2nd of November, we had excellent wind so that early on the 3rd, we saw Galveston.[46] The day before, it felt as if a big celebration had been announced. Oh, what a joy.

At sunrise we saw Galveston. Everyone rejoiced and cheered in delight to finally step on solid ground after such a long sea journey. Around eight o'clock, the anchors were cast because there was no pilot aboard yet. For there were big sand banks out there which make navigating dangerous. Around nine o'clock, the pilot came aboard to bring us close to the harbor. However, unfortunately, we were not allowed to go ashore because there was still Yellow Fever[47] in the city. For that reason we were laying anchor off-shore until the 5th. On that day at noon, we left on a steamship and arrived in Indianola[48] at noon on the 6th.[49]

By the way, I have to mention that we had bad luck with the food on our *Minna* because after 7 weeks we received only half of the bread and every evening for three weeks rice soup without broth. Our captain was quite stingy, but otherwise a good man. The other sailors were also quite nice. The captain had planned on supplies for 110 people and we were 150 on our ship. The reason for this was that he had accepted whole barges full of people on board after moving out of the harbor. In contrast, the people on the ships *Neptun*[50] and *Franziska*[51] who had left Bremen 14 days later than we did and cast anchor the same day we did, had very good food and enough bread.

We then stayed for two days in Indianola. After that our journey continued with five mules. There were four parties in our group, and the name of our wagon driver was Schlein. On the 1st day, we crossed only through prairie. On the 3rd day, we shot a wild goose for dinner, which we plucked and roasted right away. In the evening of the 4th day, a Norther[52] (a very cold north wind) with ice sprang up. We would have frozen to death, had I not had the feather beds because we had to sleep in the open air during the whole journey. On the 5th day, due to the cold weather, we took a break near a farmer's house. Then we continued our trip. Two days after we had the Norther, grasshoppers showed up and covered the whole area. There were so many of them that they almost scratched out our eyes as we were traveling on the road.[53] Finally, on the 18th [of November], we arrived in San Antonio.[54] There we stayed with Opitz from Löbau for 8 days.[55] Then we rented[56] (paid) our own quarters for 7 dollars per month.[57]

[The original letter ends here.]

2. The First Years in San Antonio, Texas: 1854 to 1857

[2.]

[San Antonio, 3 December 1856: The title page of the original letters shows a frame with drawings of images of San Antonio. On the head of the letter a picture entitled "Main Plaza, San Antonio, Texas." Underneath various kinds of leaves and fruit and in small letters the caption, "Drawn after Nature by Erhard Pentenrieder. Published by Pentenrieder & Blersch, San Antonio, Tex." On the left side of the title page in vertical order are images with subtitles of the Mission San José, Mission Conception, the Courthouse, and a man riding on a mule carrying hay. On the right side in vertical order and with subtitles are images of the Alamo, the Mission San Juan, a house with two people standing in front of it, and a cowboy, attempting to rope a buffalo with his lasso. Most interesting are the figures at the top: on the left a drawing of an Indian in proud and upright position and on the right a Mexican in the same position. In the middle of the left margin is an African American in sitting position, in the middle of the right margin, a white settler in the same position, contemplating and his chin resting in one hand, the other hand holding a rifle. Furthermore, two wild animals, on the left, a bear in a climbing position and on the right, a wildcat in an aggressive pose.][58]

Dear Father:

I need to write you again since I just worry so much about how you and my siblings might be doing. Are all of you healthy and in good spirits? By now I have written you three letters, and you have not answered any of them. In the first one I described our entire journey and also what you really wanted to know about Heuchling. My wife and I[59] do not want to be servants. Immediately after my first letter I received a letter from you quite unexpectedly, and it was a great joy for me to read that you all continue to be in good health. I then wrote you right away again, but month after month I hoped in vain for an answer from you. One Sunday, Bundesmann[60] came to me and said, 'Your father came to see my father last December and asked if he did not have any news about you.' I was quite upset about this since I could gather that you had not received any of my letters. On the third of February, I wrote another letter in which I also asked for the money because you had written to me that you had sold my house.[61] I also had put a gold dollar underneath the seal of that letter which was supposed to be the godfather gift, for Karle's Karl since he had promised me on the last day that he would write a godfather letter. Furthermore, around the same time, Bundesmann wrote to his father and took my letter along, but he, too, has not received any answer. So I would like to urge you again to send me the money, because I want to buy myself some land. I do not have enough means yet to be able to set up a farm, for it takes quite some money and effort to get a farm going. Food and rent are expensive, and one needs to have a good amount of funds before one can harvest oneself. As soon as I have the money, I want to buy quite a few cows in order to live off of them, so I hope that you will send me the money as soon as possible.

Thank God, we continue to be in good health. Karl and Christoph still attend the school in the monastery where they learn to my satisfaction. For four months now, Friedrich[62] has served on a farm; he gets six to seven dollars and food.[63] Otherwise, all three would go to school, which costs a lot of money. For each of them I have to pay one and a half dollars school tuition per month. They learn German, English and also some Spanish, but they are also in school the whole day.[64] A Lutheran priest arrived here who preaches every Sunday, which was not the case before. So now I want to have Friedrich confirmed as soon as it works out. In the beginning, we were afraid, for there was no mention that there was a protestant church service available, rather we heard things like 'today someone was shot or stabbed to death.' Even at dances there are often shooters ......so the shooter is the third or fourth man."[65] For here a human life does not count more than the life of a dog in Germany since strict investigation is not done, especially if it concerns the lost life of a German or Irishman.

This year I spent a few months mowing in the prairie again. I get a dollar per day and food for doing this.[66] Later I mowed for myself and made a haystack, which I want to sell this winter.[67] My wife washes and irons for the people here, including Bundesmann and other store clerks.[68] She gets one dollar for the dozen and has already made good money for herself this summer. This year was a very bad one; the winter was so cold that none of the Mexicans here could remember having seen ice that thick before. Equally, the summer was very hot so that the corn[69] and sweet potatoes[70] did not turn out well. German potatoes[71] do not grow in the San Antonio area, because the soil is too heavy. They grow best in sandy soil. Three pounds of potatoes cost 25 cents, 12 pounds wheat flour one dollar, corn flour 1 dollar 30 cents. Thus, everything is quite expensive this year.

Now, dear father, I seriously plan to buy land. Even if I have to pay a little more per acre, I want to see to it that I have wood and water and do not get too close to the area that borders the Indians' land.[72] Actually, they think they do God a favor if they can kill a white man. They also like to steal horses, mules, oxen, and cows.[73] I now own 5 cattle, 2 cows and three calves, two of which will have new calves next year. I also own 20 chickens. In San Antonio they build more than 150 new houses on average every year. Still, there is little work for me available this winter for there are only carpenters, bricklayers, and merchants here and when there is work the Polish get it because they work for less than the Germans. So I am determined to farm.

Dear brother-in-law, you asked me in your letter for flower seeds, but, unfortunately, this is something very difficult to get here in the San Antonio area. The reason for this is there are way too many cattle here which eat the flowers. Once I live further away from the city, it will be easier to get. There are few fruit trees here, but we do have pear, apple, plum, and a few cherry trees, although there are wild kinds of all sorts. Peach trees have been the most profitable ones so far because, if one puts peach stones in the ground, they very often produce fruit already in the second year.[74] I planted 2,000 peach stones for myself. I also want to put out some apple seeds. My landlord gave me a lot in the yard for the seeding. Last year, I also slaughtered quite a few pigs and have started to do so this year.

I would like to ask you, dear father and siblings, to write me about everything you are doing. August, do you not have children yet? And Luise, have you not yet gotten married? If this is not the case, write me so I can also celebrate your wedding here in Texas. Admittedly, I cannot lead the calf for you.[75]

In San Antonio, the Germans act in plays.[76] Bundesmann, a man from Löbau, and Becker participate in them. Usually, Bundesmann plays the best parts. August, that would be something for you; you could play the part of the tomfool. Please do me a favor and say hello to my old neighbors, especially Wendebaum and the cartwright, but also my "flower colleagues," especially Neumann as well as Karl Tischler, Obstel, and Langkreißel. I also should mention that I would like to ask you to notify my godfather Bergmann in the Haine[77] of how I am doing.

Dear father, when you send me the money, you might have to purchase a New York bill of exchange since the merchants here make most of their transactions with them. Also, I talked with Bundesmann today and told him that he might want to pay out the money to me as soon as he has received the letter from his father, should there be too many difficulties with the bill of exchange. He, too, planned on writing to his father tonight to ask him to write back as soon as he has received the information or the money. Dear father, please take care of this matter. I will need a notarized statement in any case so I can show Bundesmann if you have paid out the money to Bundesmann senior or if he wanted it as a mortgage.

Please say hello for me to all my friends whom I cannot all name, and tell them that I am doing fine. Admittedly, Texas will not be a Germany. Although I do not have to work very much, there always are also many worries. For when a month is over we must pay 5 dollars rent, although we only have a small room and it is still considered cheap. Work is paid well, from 75 cents to 1–2 dollars, but what costs five German pennies for you is more likely to cost 5 cents here.[78]

I also have to mention that there are quite a few thieves and bad people, and also drunkards here (called *Laufers*),[79] who will hear nothing of work. The city is now rather big and has all kinds of people from all imaginable countries. Even Arabs with their camels are not missing.[80] Many Germans left Germany because they committed fraud and they engage in similar matters here because the Americans do the same. You are not able to trust anyone. Money earns fairly good interest. The ones who have any get at least 2 dollars per one hundred dollars a month. The most peaceful and best life in Texas is that of a farmer, and it is also the cheapest because cattle thrive here and do not need any feed, nor do they require much work. It is especially so for the pigs since they multiply in 1½ to 2 years from 6 to 8 mother sows, so much that one can sell them for 4 to 500 dollars.

I have to admit that we did not like our first year here at all. Now things are going a little better, but I still see the last evening with you in my mind, how I said farewell, and I get choked with emotion when I think of it. In my dreams I have often talked with you, but realized that it was just a dream when I woke up. Our English skills only progress slowly. We can understand more than we can say, so Karl or Christoph often have to speak in our place.[81]

We have heard a number of things about your area, even about Ebersbach itself, through the Löbau newspaper that Bundesmann lets us read. We are geographically about 1,800 miles from you, and the day differs by 9 hours. When you eat lunch I am still fast asleep as it is three o'clock in the morning. Our daylight hours are now from six o'clock in the morning until six o'clock in the evening. In the summer the day lasts from four o'clock in the morning until eight o'clock in the evening. Daylight comes suddenly, and so does nightfall. There is no such thing as dawn or dusk here. If you set up a pole of 6 feet so it is vertical, it casts a shadow of 11 inches on the longest days. Forgive me that my handwriting is so poor, but we have forgotten how to write, and the quill was also not the best. There is one more thing I must ask you: let me know how my brother is doing and how his children are; probably they do not give him much joy, but tell me about him either way, be it bad or good.

Dear Grandfather:
In order for you to see my handwriting again, I am adding a few lines and wish you much happiness for the New Year. Should the letter not arrive on time, accept my wishes anyway. I continue to be healthy and work in a beer hall in the city, across from the church, which you see in the picture. Please say hello to my teacher Mr. Kantor for me and tell him that I have very much enjoyed the *Liederkranz*[82] that he had given me as a farewell gift, especially the "Trip to the Blueberries."[83]
Your oldest grandson Ch. F. Bergmann

Most beloved Grandfather and Godfather:
I also need to write you a few lines so you will see that I still am alive. I go to school now from early in the morning to evening and have to study a lot.
Many greetings to my teacher, and tell him that I have to study much harder here than in Germany.
I remain,
Your second grandson who loves you very much.
K.H. Bergmann

Dearest Grandfather!
That I still am alive you see from my signature,
Christoph Wilhelm Bergmann

Many greetings from your loving son
Christian Friedrich Bergmann
Joh. Christ. Luise Bergmann, born Dreβler
San Antonio, 3 December 1856

[3.1]
San Antonio de Bexar, 2 June 1857

Dear Father and Brother, Sisters and all Friends:
I feel the need to write you another letter. I had meant to write you long ago, but I wanted to be sure I could give you exact information about my land, how much I own and where it is. Also, I was sick for three weeks, and Friedrich

also fell ill during that time with a case of scarlet fever. Then my wife had a bout of intermittent fever[84] that also lasted for three weeks, but we did not have a doctor.[85]

I own 320 acres of land in Comal County,[86] right by the Guadalupe River,[87] on which I have two houses but not much fence yet. It cost me 200 dollars.[88] I wanted to move there this spring, but the great drought prevented me from doing so, as it did not rain from mid September to mid May. In the evening of Easter day, there was just a little rain mixed with hail, followed by such cold weather that at eight o'clock in the morning, there was still a thick layer of ice on the roads. This had never happened before. But twice in the past, the corn had frozen, which caused a fairly steep increase in prices. A bushel (1 Bushel = 36.35 liters)[89] corn (*Mais*) costs one dollar and eighty cents, and that is rather much, if one has bad earnings.[90] Next fall, however, I want to move to my new land.

Friedrich and Karl were confirmed this Easter.[91] Friedrich is still working in the beer hall, and Karl works in a store, but does not earn much money yet, only 4 dollars a month. However, he is supposed to make more down the road. I do not want to take him away from this job since he is not that well suited for farm work. That is the reason why I want him to learn to become a merchant. He has become rather good at reading and writing in English, and I believe he has no difficulties with learning.

I also should mention that they are building a silver mine two hours away from my land, and they have found silver. In the area around, and also in San Antonio itself, there are a lot of thefts and murder. Within a short time period, there were repeated burglaries. Even travelers were murdered and robbed. Horses, mules, oxen, cows, etc. were stolen, but now things seem to have changed. Last Friday morning, three of the men who were accused of robbery and gambling were shot to death, several others were hanged and shot to death earlier, and the plan is to continue to proceed like this. All those who are known to have committed robbery and murder or are suspects are supposed to turn themselves in by mid March. If they do not they will just be either shot to death or hung, either way they die. Furthermore, a sheriff (police officer) was shot to death, and one citizen was severely wounded. Several people were arrested and found guilty.

Last winter in January, I slaughtered some pigs. Later, I grafted trees which, lately, brought me 2 dollars per day. I will do this again in the beginning of the next month. After that, I moved 2 miles away from the city into the country side. However, within 3 weeks, due to the drought and wrong expectations, I moved back into the city where immediately I found work again. There I also learned how to lay bricks. Of course, it is easier to do than in Germany because the stones are either cut with a saw or hewn with an ax. Later, I would like to get hay in again.

Dear father and siblings, your letter, which I received through Bundesmann on March 14, came quite unexpectedly and was all the more delightful since I also had the joy of reading your handwriting again, and the money arrived exactly at the right time. I had borrowed 125 dollars from Bundesmann for which I had to pay 3 dollars and 25 cents interest per month,[92] because I had not sold my hay yet and therefore had run out of money. Bundesmann paid me the money in cash in the correct amount and in full, and I signed a receipt and think that you probably already hold it in your hand. My wife still works in her usual job. She does laundry and now and then some sewing work. Christoph attends school in the mornings and in the afternoons and recently has practiced reading, writing, and arithmetic quite a bit.

Dear brother, you had written me that the stork had so much work to do. Was it not also up on the barn? After all, a stork also likes to stay on farms, and it was already quite busy when I was still there. So far I have not noticed any here yet. Perhaps it feels it is too far. I do not think I will see birds like these here.

Also, dear sister, I could not be present at your wedding, but I wish you with all my heart happiness in your new and different life and that you will enjoy a peaceful and blissful life with your husband. God willing, my letter finds you in as good health as yours found me vice versa. I would like to ask you to write me again soon. Give my regards to all of my acquaintances and friends. We continue to be in good health. If we get some more rain again, the harvest should be rather good, and things soon will pick up again in Texas. I will definitely move to my land next December, because it is and will always be the best land in Texas. I will write again shortly since two people from Reichenberg who live in our city will return to Germany[93] and are willing to take my letter along with them.

I remain,
Your son who continues to love you,
Christian Friedrich Bergmann
San Antonio de Bexar, 2 June 1857

[3.2]
[Texas, San Antonio de Bexar, 12 June 1857]

Dear Grandfather:
I need to write to you, also on behalf of my brothers, so you can see that we think of you and speak of you all the time. I am healthy again after I was very ill and did not expect to recover. At Easter, my brother Karl and I[94] enjoyed the Holy Communion for the first time because there is a protestant preacher here. His name is Zitzelmann, and he is from Württemberg. I now work in a beer hall and get seven dollars per month because the times are bad. Otherwise, I would receive 10 dollars.

For the time being, we want to stay in San Antonio, but in the fall we want to move to our land in the mountains,[95] 35 miles further north. It is healthier to live there than in the city. Not far from there, I drove oxen last year and I liked the area very much. Currently, there is a great deal of unrest in the city because they have discovered a gang of robbers that have made all of Texas unsafe. Often the Indians were blamed in their stead, but Indians were rarely involved. And now they have started to rob in the city. They murdered 1 farmer on a road not far from the city. A secret society has formed whose members go after them and wherever they find any of the gang members they hang them right away from a tree or shoot them to death, like they did in mid-May. Back then, three were shot to death in the street, among them a gang leader. The 11th of this month, 11 men were hanged, 7 from a tree and 3 from a tree a few miles from the city, and one, who admitted to everything, from a tree right outside of the city. That gives you an idea of what is going on here. One more thing I have to mention: there was only one German among the gang members. All the others were Americans, Irishmen and Mexicans.

Many greetings to my friends and Karl Tischer. Tell him he should be so kind to say hello to all my comrades. He knows them all, especially August Gochtliebel and Mr. Kantor. And do write us again soon, so that we know what you are up to. Forgive my poor handwriting, but it is evening and I wrote in a hurry. Farewell; think of us often.

I remain,
Your loving grandson,
Christian Friedrich Bergmann
Texas, San Antonio de Bexar, June 12, 1857

3. Farm Life on the Guadalupe River: 1857 to 1888

[4.]
Blanco County, 16 January 1859

Dear Father and Siblings and also Brother-in-Law:
I am sure you have waited for an answer for a long, long time, but I only received your letter in December because we had not been to San Antonio in a long time. I went to the post office several times, but there was never a letter for me. In December, I returned to the city and went to see an acquaintance who had picked up the letter from the post office because it had been mentioned in the newspaper, and he gave it to me. Please do not send the next letter to San Antonio again, but to the address at the end of this letter.

Dear brother-in-law, in your letter I read with great interest that you all continue to be hale and hearty and also that Luise got married. We wholeheartedly all wish her happiness. As for you, dear brother, we wish you luck with farming. Dear brother-in-law, you wrote that father did not want to believe the oral report that Bundesmann gave about how we are doing. That is why I want to begin to tell you more. I have 320 acres of land of which between 20 and 21 acres are fenced in. I have built a stone fence 700 feet in length, the remainder is a wooden fence.[96] The land is located at a large river, called the Guadalupe, and, therefore, I have very good water and also enough fish, if we have time [to catch them]. There is also plenty of big game.[97] However, so far we have not had the time to shoot any. But we always have meat to eat because when I happen to shoot a deer we have food for a while, or a turkey (*Truthahn*), of which we have an abundance here. They also make excellent soup. Some of them weigh 24 to 26 pounds. Dear father, I now want to list my complete livestock for you: I have 3 yokes of trek oxen, so altogether 6; 6 cows and 8 calves, one horse, four pigs, and three dogs. One of the dogs is five weeks old and cost me 3 dollars, and one is four weeks old and cost 2 dollars. They are English Bull Dogs who do not put up with foolery.[98] I do not have many chickens yet, only close to 20.

As far as this year's harvest is concerned, it was not that great for me because I could only plant when it was basically too late in the season. The reason for this was that first, I suffered from dysentery, and second, one of my oxen for which we had looked for a long time died. There was not much we could do about it since we had no money and no one knows us here. But I did go to an American who lives three miles away from us. I borrowed a wild ox from him and paid it off within a month. It cost me 18 dollars. I believe that, God willing, I will have a better harvest this year. On two acres I planted wheat which, if it grows well, will make 40 bushels. 3 bushels equal about one Dresden Bushel.[99] One bushel of wheat weighs 60 pounds. I also want to plant cotton and sugar cane this year. These items are quite profitable. I admit growing cotton involves a great amount of work, but we are enough people to take care of it.[100] I also have big cypress trees, from which shingles are made.[101] When I do not have any other work, I make shingles, for they are easier to make than in Germany because one does not groove them. One mounts them like roof tiles in Germany, and they are sold by the thousand. If the price is good, they sell for four to five dollars so that I always have a good day's wage. When I will be finished with plowing, Friedrich will have to go to San Antonio to sell cedar wood for fencing. If the price for the wood is good, I get 8 to 10 dollars for 100 pieces. With that money one can buy coffee, sugar, and other small items to take home.

Dear brother-in-law, you wrote that you often dreamt of us. That also has happened to me and us, but I always dreamt that I was annoyed that I had left my property [in Germany]. Most likely, I will not return to Germany, for I could not get as far ahead there as I can here. My cattle are growing and, as soon as I can, I will buy even more cows because it does not matter whether I have 10 or 30 cows. I have to look after them twice a week anyway. I can honestly say that my cattle do not cause much work because every time they come home, my wife gives them salt to lick and, if she does that regularly, the cattle do not stray far from the farm. By the way, thank God, I have not lost any cattle yet.

Not far from my place is a grist mill close to which is also a saw mill where I can have various wood cut into boards and posts and then take it to town. Of all my income, I pay 4 and ½ dollars in taxes each year, while even the smallest house has higher fees. It is indeed true that whoever has not learned a trade should go to Texas. I am everything here, a shoe maker, a wagon maker, a cold smith, a butcher. In short, I do everything that needs to be done. I even geld the oxen and pigs myself and also have to do that for the neighbors.

As for the rest, in wild fruit we have cherries and plums, and early and late wines.[102] I have not made any wine yet this year, but next year I also want to make some wine. There are farmers who made 10 to 14 buckets[103] of wine. I just have not really had the courage yet to try doing it. I have about 1,500 peach trees, 50 of which produce quite large fruit. Here in Texas the peach trees bear fruit as early as their third year. I also have some green plum trees and apricots. Now I just wished I had some of the kinds of cherry trees [that I had in Germany]. Apple and pear trees I am able to get here.

I want to finish my letter now. We are all healthy and wish that this letter, too, may find you in good health. Please write very soon. I admit I could have written a month earlier, but there was always something that kept me from doing so, and that is the reason why it took me so long. So please do not hold it against me. My address is Mr. C.F. Bergmann, Hodges Mill,[104] Blanco County, Texas.
I remain,
Your loving son
Christian Friedrich Bergmann

Dear Brother:
Say hello for me to all my acquaintances and friends, and especially Wendebaum, the wagon maker Weber, and Bergmann in the Haine. Please let them read this letter since they asked me to write to them. Please say hello to my flower colleagues,[105] too. I wish all of them happiness and health, and the flowers I have forgotten. Now and then it still hurts a little.
Dear Father, send your letter via Bremen and New York.

[5.]
[Blanco County, 1 December 1859]

Dear Father:
I received your letter from which I could see that you all continue to be healthy. Dear Father, you mentioned in your letter that I did not write you often enough about my land. I had a lot of land fenced, but did not have much cleared, only 15 acres, of which on 9 acres I planted corn and on 3 acres cotton. On the 2 acres on which I had planted wheat, I also planted corn on July 4, after I had harvested the wheat, assuming that it would be good feed for the oxen. But the weather was favorable, so I ended up getting a good harvest of corn (corn is the name for *Mais*). I also sold melons for 29 dollars, which I had planted in between the cotton plants. There were four hundred of them. I had grown many more, but they did not bring in good money in San Antonio.[106] That is the reason why I fed them to the pigs of which I now own 27. I intend to butcher three of them and also an old trek ox because he always jumps over the fence and has already caused me a lot of damage. It is not enough that he crosses over alone, but also the other cattle follow him and get used to doing so.

This year's harvest was significantly better than last year's. I think I will harvest two bales of cotton, each of which weighs 500 pounds and sells for 50 dollars. I harvested 120 bushels of corn, of which I want to sell 70 bushels. I also want to mention that I planted Chinese sugar cane, from which one makes syrup. I just tried it out this year, but next year I want to plant more since I could already make 80 canisters[107] of syrup this time. Sugar cane grows 8 to 10 feet in height and produces many seeds that I feed to the pigs. I also built the sugar cane press myself, which consists of two rollers and is turned by a horse. That was hard work for me since I could not use a lathe to make the rollers round, only a planer, and I had a difficult time to give them a round shape. I also harvested 450 pumpkins. We eat and sell only the nicest ones of them, the others I just feed to the pigs because, when cooked, the pumpkins make a good mash. This year, I also experimented with sweet potatoes. It was only a small trial patch because I did not have any seedlings of my own, and it was hard indeed to get any at all. But I have grown quite nice ones as well although I do not know yet how many. I have not figured it out yet since I am still busy picking cotton.[108]

This year, I planted 2 acres of winter wheat seeds. On 2 acres I want to plant rye seeds and on another two acres summer wheat seeds.[109] The threshing of the wheat troubled me greatly. (trouble = Mühe "getrubelt" = *Mühe bereiten*). I had no flail,[110] no barn, I could only thresh in the house. I made the flail-cap and the connecting belts from rawhide, the flail and handle I made with a plane. Making the handle gave me the hardest time, but I managed to succeed.[111] I also want to plant half an acre of tobacco this year. I would probably plant more but I am not experienced enough in growing tobacco.[112] My neighbor planted 1½ acres and harvested over 2,000 pounds. But he also had a man with expertise in the area. The profit for the crop will amount to 250 dollars.

I am currently building my house, which I am making from cedar wood. The cedar wood here is the same as the wood of which the garden fence in Friedersdorf[113] is made. The house will be 48 feet long and have a hall and in the front a porch. It will be 36 feet wide and have a breezeway in the middle with a width of 12 feet. The construction method of the house is as follows: one puts tree trunks on top of each other and then latches the ends together so that they do not rest upon each other. The spaces in between are filled with stones and clay and the tree trunks trimmed on the inside and the outside.[114] When I have finished the house, I can thresh in the breezeway.

Dear father, you also wanted to know what we do with the butter and the milk. Here a cow does not give nine to ten pounds of butter like in Germany.

The calves suckle until they are nine months old and often even longer, depending on when the cows have calves again. One may take only half of the milk because calves may not be outside of the pen[115] (a pen is a piece of land that is fenced in with rags of fabric). One does not feed anything but salt to the cows, and they do not expect to get anything else since they are not used to it. And so I do not worry about what I should do with the butter and the milk. When we have enough I am happy. If mother saw how much butter we eat, she would say: "You are stuffing yourselves."[116] For the same reason, our neighbors always say that we take this matter very seriously. However, our climate requires that we eat a good amount of fatty food; otherwise it would be hard to do all the work.

I also have to mention that there was a disease amongst the cattle and the pigs, even deer were affected by it. I, too, lost 3 sows and a one-year-old ox, and one calf went blind. But the loss does not hurt as much here as it would in Germany.

Dear brother-in-law, you wrote me that Freude wanted to know about how Heuchling was. I spoke with people from Castroville on my way to San Antonio, and they said that he first earned a nice amount of money making music, but that this was over now and that he has gotten somewhat out of the habit of working. I also talked with the daughter of Feine's brother (at the time she was employed at Marx's garden center in San Antonio), who told me that he was living in constant strife with his brother. I myself have not yet talked with him. He was in San Antonio but did not come to see me. However, I know this much: he is not in good shape.

Dear father and brother, I also have to bother you with a request: I intend to have a new wagon built for myself, namely by Gottlieb Weber and Ulrich in Ebersbach. However, it needs to be so strong that I can load 60 centners[117] and drive over hill and dale. Yet I want only the four wheels and the two shanks. But the shanks must be of iron and must not protrude more than two inches because the roads here are quite narrow in places, due to the trees. Also the wheels may not have a camber since one should not take things lightly here. For when the roads are bad, the wheels can sink so deep that the axles touch the ground. But I also do not want to have bright white spokes in the wheels. Actually, I just need good wood, and for that reason Weber should build the wagon. The front wheels need to be 2 cubits and 7 inches high and the rear wheels 2 cubits and 19 inches. Dear father, make sure you order the wagon right after you have received my letter so that the wood will dry well and I will get the wagon by next fall. I hope that you will lend me the money for the wagon-maker because I have not yet sold any of my harvest. Please talk with Bundesmann about the payment. He can tell you what the best way for me is to make the transaction, whether as a note or in cash. But write me as soon as possible, so that I know how expensive the wagon will be or whether you do not want to deal with all of this. I forgot to mention how wide the wheel track of the wagon must be. It has an inside diameter of 5 feet and 2 inches and has no shaft and no shear hook, for the shaft of a wagon pulled by an ox needs to be able to move up and down, and I want to work on this myself. But I still would need a good and functional screw wrench in order to be able to tighten smaller and bigger bolts.

I have now four yokes of oxen. I replaced the one that died with one that cost 23 dollars and also bought another young ox to add to the one I have raised myself. That one I got cheaper, for 8 dollars. I hope that by next fall, I will have five yokes of oxen because I have another young one and intend to buy an additional one.This completes my letter, and I wish you that it will find you in good health.

I remain,
Your loving son, brother, and brother-in-law
Christian Friedrich Bergmann
Blanco County, 1 December 1859

[6.1]
[San Antonio, 3 March, no year, probably 1861]

Dear Father and Siblings:
I received the letter from Richter and was very glad to hear that you all continue to be healthy. However, I received the letter only on February 3, because we had not made it to San Antonio since Christmas, and Richter did not take it to the post office. Only when I came to see him did he give it to me. Richter had a good trip and arrived in San Antonio in good health and good spirits with his father and his sister.

Dear father and siblings, we are going through quite alarming times here in America. The southern states want to separate from the northern states, and indeed eight of them have already seceded from the North. The separation is merely due to the new governor, because a new one was elected in the fall. The North is against slavery and the South is in favor of it. Now the North succeeded in electing a governor who is against enslaving Negroes,[118] and the South is outraged about it because it has many slaveholders. For that reason it has come to the separation. These circumstances are a big setback for Texas which, of course, takes sides with the South. In a short time, the Government of Texas will be removed, as will the soldiers who are stationed at the borders of Indian Territory. This has already had a big impact on San Antonio in that the prices of all raw products have significantly fallen.[119]

Dear father, the fact the Richter did not bring along the wagon has been a big loss for me because I had a lot to haul, and my old wagon is fragile and basically ruined. If I wanted to have it repaired, it would cost almost as much as a new one. So, right now I can only transport little to nothing and have to let my oxen run around unused. Of course, I have my fields in good shape and have also already fenced in more of my new land because Friedrich is almost as strong as I, and Karl not much less. Christoph is with a neighbor where he drives oxen and also does other work for which he earns 5 dollars a month.[120] The people are very content with his work.

Dear brother-in-law, you asked me in your letter whether we have not also seen storks here. We do not know such birds here, and I do not believe that I will ever see any again. We grow old and do not worry much anymore about such things.[121]

Dear father, I regret that I agreed to order a second wagon since my neighbor Wetzel has not even asked about it yet, and also I am at odds with him. I fenced in five acres of land which bordered my fence and had intended to buy this land later, but Wenzel went to the surveyor and had it surveyed without my knowledge. Of course, this has greatly annoyed me, and I have not gone to his house since then, nor has he come to see me. However, do send the wagon because a different neighbor wants to have it.

Dear brother-in-law, you have not written me at all about my old comrades and flower friends.[122] I would like to know how they are doing. Please also say hello to my godfather Bergmann in the Haine. Tell him that we continue to be in good health and that I can barely remember the flowers anymore because I have slowly forgotten them, but that I have planted a fairly large orchard with apples, pears, plums, cherries, peaches, and so on. I did receive the seeds together with the yellow plum seeds and was very happy about that.

I assume you will pack the shanks in a box. Please pad them with some oakum because we badly need it to clean our guns and cannot get it here. I am sending you a note for sixty-four thalers[123] and 28 *ngr*[124] and am sorry that I cannot pay for one wagon in full. I just could not quite get all the money together. I had to sell three cows on credit, but I think that I will be able to pay the rest of the money by the time I receive the bill. Insure the wagons, and as for the shipping to Bremen or Hamburg, just put it on the bill. Please send the wagons to A. Runge and Co., Indianola, Texas.

Farewell, and I wish you that you will be in good health when you hold this letter in your hands. We are all healthy and send greetings to all of you.
Your loving son
Christian Friedrich Bergmann
Blanco County, Hodges Mills

Enclosed is a dollar for Cartwright Weber. It is worth 1 thaler and 1 ounce.[125]

[6.2]
[Loose note in the correspondence, no date]

Dear Father:
In case you have not yet sent my money that is still in your hands or have not yet paid it to H. Bundesmann in Neugersdorf, I would like to ask you to send him 150 Prussian thalers[126] right away. The remainder should be sent to me through him by promissory note to New York, or send the entire amount by money order to E. Bundesmann here in San Antonio for Mr. Bundesmann in Neugersdorf. As soon as I have the money order, I can get the money from his son.
It is all the more important to me that you take care of this matter as soon as possible since I already have received an advance on the money.
Your loving son
Christian Friedrich Bergmann
San Antonio, March 3

[7.]
Guadalupe, 1 July 1865

Dear Father, Brother, etc.:
Finally I have a chance to write you again. It has been 4 years since I have heard anything from you.[127] From your last letter I saw that all of you continue to be in good health and that has been a great joy for me. Dear brother, you wrote about the wagon and whether you should send it via New York. But when I received your letter, the port was already blockaded[128] and I could not write you an answer. I received the bill for these wagons and, in my estimation, it is not too high. I bought a fairly sturdy wagon, also with axles made of oak, onto which I can load 60 centners. I paid 330 dollars for it. Of course, it was a great loss for me that the wagons could not get here because there was much money to be earned. However, I believe and hope that I can still receive the

wagons, and I also need them, although not right away. For there will be a time when each Friedrich and Karl also want to have wagons, and I intend to give these to them. I would have put the remainder of the money for the wagons in my letter, but the postal service is still too unsafe to send promissory notes. Dear brother, please write me what is going on with the wagons. In case a ship leaves before I can write again, just send them, namely to Mr. Runge & Compagnie in Powder Horn. As soon as I have an answer from you, I will send you the money for the wagons and possibly the gold dollar that you requested, even though I have already sent a letter with 3 gold dollars, but the letter must have gotten lost.

Dear brother, you sent me a listing of your entire livestock, and you also wanted to know about mine: To begin, I have 3 riding horses, one mule on which one can also ride, 2 old mares, two two-year-old mares, 2 one-year-old mares, two two-year-old geldings, 4 yearling stallions and one colt, 16 trek oxen, 31 cows, 11 three-year-old wild oxen, 14 heifers, 24 one-year-old cattle, 16 calves, therefore fewer than last year.[129] 12 cows died (there was a disease among the cattle), and the others are not producing calves. I have only few pigs, 19 of them, and 44 sheep, but they are not mine. They belong to Karl and Christoph. I also have 30 chickens, 3 pairs of doves, 7 geese and 20 ducks. That is my entire livestock.

I cannot write much about politics for, if I wanted to, I would never be able to finish. All the boys got off with barely a scratch. Friedrich had a nervous condition, but had to drive for the Government for a while, Karl was always not old enough, but they have grown quite a bit.[130] Friedrich is as tall as I am, Karl is one and a half inches taller, and Christoph will also soon be as tall as I. At any rate, we can all wear the same shoes. By the way, we are all healthy and wish that our letter finds you in good health and good spirits.
Cordial greetings from all of us
Your son and brother
Christian Friedrich Bergmann

Address: C.F. Bergmann
Boerne, Kendall County, Texas[131]

[A loose note accompanying the letter]
I need to have these lines returned. Dear father, I have received three hundred and ninety dollars from Bundesmann.
Christian Friedrich Bergmann

[8.1]
Guadalupe, 28 June 1866

It was Monday, June 25. We just had our county and state elections in Boerne when I rode to the voting station with Friedrich and his father-in-law in order to cast our ballots. At the same time, we also wanted to get the newspapers since it was mail day. For that reason we stayed there for another few hours until the mail arrived. I then sent Friedrich immediately to get the newspapers and, to my delight, he also brought back your letter which I opened and read right away. I was overjoyed to read that you continue to be in good health, except for the big one[132] and his wife and Christliese. Then Friedrich discovered the three pictures of you, dear father, which at once went through many hands. The merchant where we were, immediately cut the glass for the three pictures because, as he put it, the face was too endearing to not make the glass for the frames for them. Afterwards we also drank a couple of beers to your health and then rode back home at full gallop, where everyone had been asleep because the trip home was 15 miles. But when they heard that we had a letter from Germany, they all became as awake as a flock of flushed birds. You probably can imagine what a great joy it was for everyone. My wife, dear father, read your letter with tears in her eyes, and everyone was just so happy that you were still alive and in good health, and especially to see the pictures of the grandfather.

Dear father and siblings, etc., my farming life has indeed changed considerably. Friedrich married the daughter of a neighboring farmer.[133] They came here from Upper Silesia. Karl is also engaged, but he wants to wait for another year before he gets married. His girl is more or less the foster daughter of a German man for whom her mother does the housekeeping. He is unmarried, and that is exactly the main reason why he is for the marriage. He will not get married himself and no longer wants to have to take care of everything alone. So he told me—not knowing how much longer he would live—Karl should take over the farm.

Dear father, you wanted to know what kind of house I have, so I am sending you the floor plan. I have a friend who can draw quite well. One of these days when he comes to my house, I want to have him make a drawing of it so I can send it to you. I assure you it is the most beautiful and best house near and far, but it has cost me much work and money, too, and so it happens that it is the object of envy for some people. Recently, a man visited me, and my wife urged him to bring along his wife the next time he would come over. Yet, he said that he had better not do that. Otherwise, his wife would give him a hard time if she saw our house.

Dear father, you also wanted to know what kind of wood grows on my land. There are four different kinds of oak trees, live oak, post oak, black oak, and white oak. But I do not think that they are as strong as the ones in Germany. I also have other kinds of wood by the Guadalupe River which I use as lumber wood, the red elm, the schlicker elm and the sycamore tree. Right by the water, I also have cypress trees from which we make shingles that are quite strong. This spring, I cut one that measured five feet in diameter, and my neighbor cut one that measured seven feet in diameter. On the other piece of land there are one hundred acres of cedar woods (Duja). I drive the wood to San Antonio, but it is also quite good for making fence posts. Then there are another twenty-eight acres of various kinds of lumber wood, bois d'arc,[134] black walnut and hackberry trees, also some linden trees; there are no pine trees, spruces, larches, alders and birch trees here in West Texas. In East Texas there are pine trees. I should also mention how large one acre is.[135] An acre is 75 yards long and equally wide. I have 608 acres total, but in two pieces. The cedar forest is 3 miles away from my house.[136]

You also wanted to know what generally grows on my land. All fruit does really well if the weather is favorable, but sometimes there is no rain so they do not turn out well. It all depends on how lucky one is with the plants. This year, my harvest will be even better than expected. The main crop will be corn; I will not have wheat and rye this year. I grow wheat and rye, peas, sugar cane for syrup, sweet potatoes and millet. Millet is a kind of grass which I need for the horses in the winter. All these plants grow well here. As far as the orchard trees are concerned, peach trees do best, but apple and pear trees do well, too.[137] Last spring, I planted quite a few peach trees. They usually carry fruit as early as their second year and grow at a phenomenal speed. We can grow basically any kind of vegetables here. Last fall, I grew lettuce, which we ate until April. I have a nice garden for vegetables and trees and even have a spring there, but one cannot use it as a well. I have to let the water run for a short distance through channels, from where I collect it in a 10-ton trough that I made from cypress wood. Next fall, I also want to plant wheat and rye again.

However, since my last letter, I have also had quite a few losses: 3 yoke of trek oxen and 3 cows, but the damage has been repaired. I was especially lucky with the horses. I have 4 riding horses and one mule. One horse I gave to Friedrich. By the way, Karl and Friedrich each have 2 horses, Christoph has 1 horse. I also have another 11 horses which I use for breeding. Among them are 2 tame mares. Now I would just need to have a good stallion, but I have too few horses because a good stallion costs 400–500 dollars.

Dear brother, you mentioned in your letter that I wrote so little. That might be true, but if I think of the past I get really disgusted. Actually, I would not find an end if I started. We made it through everything as well as we could to not become soldiers.[138] In the past few years, so many atrocities were committed as probably had never even occurred during the French Revolution. Those of the opposition party[139] tried several times to make us soldiers, but in vain. They could not conscript us, and we did not volunteer to go.[140] In fact, my house is in a location that is ideal for escaping. In the last year of the war, I had many visitors, namely deserters. Woltersdorf from San Antonio and his family stayed with me for a full year. Woltersdorf is the son-in-law of Richter Senior of Groβhennersdorf. Next fall, Woltersdorf wants to go back to Löbau, but the way things look where you are, I think it will not happen. He promised me to visit with you when he gets there. Then you will hear from him in person how things are going for me.

Dear father, I intend to take care of our matter through Hermann Woltersdorf since that way, when he goes to Germany, it will cause the least problems. I hope this letter will find you in good health.

I remain,
Your loving son, daughter, and grandsons
Christian Friedrich Bergmann
Johanna Christiane Luise Bergmann
[Included in this letter was a drawing of the floorplan of Bergmann's house.]

[8.2]
[This letter was most likely sent with the previous one since it was written on the same light blue stationery that Bergmann rarely used, and the handwriting is consistent with the one in the previous letter dated "Guadalupe, 28 June 1866."]

Dear Brother:
You would like to know whether you should come to Texas or not. That is a difficult question for me to answer.[141] I have heard from many people that they were often blamed of having offered bad advice in this respect, and I would not like this to happen to me. For, when I think of us and when we came to this country, if someone had been responsible for our coming here, I would have blamed that person bitterly for encouraging us. Therefore, I do not advise you to come here, but I also do not tell you that you should stay away and

worry for your whole life about debts. It is a whole different situation for you than it was for me. I can help you with my advice and in deeds. This was not the case when I came here. You would also have friends here that I did not have when I arrived. The social life here in the country is not like in Germany. The houses are too far apart, and you are not able to just talk with your neighbors.[142] As a result, one easily gets homesick. If this is not a big issue with you then I would say, come to Texas. There is still much good land for sale. Certainly, you will not be delighted at first when you see Texas, just like I was not. But if you show good will and are seriously determined, your eyes will get used to it. If you do want to come, there would be a good piece of land for sale just 3 miles away from me. The owner recently visited here and I told him that you were thinking of coming to Texas. He assured me that he would not sell the land, unless he knew for sure that you would not come. But you could also lease land for one or two years and get a piece of land ready for yourself. If you are in good health, it is a way to earn some money. My wife, my children, and I do not want to go back to Germany, at least not in order to live there, however, perhaps for a visit.[143] This year, I paid 6 and ½ dollars for state-, county-, and poll-tax. Those are all the fees because I was in the 19 hundred [dollars] tax bracket; at any rate, it was the lowest bracket that I could get since the tax assessor determines the amount for my district.

Dear brother, if you come, make sure you make yourself as comfortable as possible. Sell everything you can sell. Since you also inquired about what you should bring along, have someone bake you zwieback,[144] [bring] dried fruit, plums and apples, 4–6 dozens, real Vienna balm,[145] and 2–3 pounds Hoffmann's drops[146] because the latter are very expensive here. Also bring Atlas linen[147] for pants and shirts, Dresden cloth[148] for ladies skirts, the whole set of a centre-bit,[149] some good wide hoes, but not round ones, 3 harmonicas, about 10–15 guilders a piece, from Schönlinde or wherever you can get good ones. Also, buy a good double-barrel shotgun.[150] If you still own tight-fitting clothes, sell them all because these are not the style here and are also too constricting. Everything here has to be loose-fitting. Even the pants would not be worth a thing, and one does not have any use for suspenders here either. Also bring other color fabric for pants so one can make work pants from it. Bring me 2 dozens of hand-towels, six quarter in length. Also, since one cannot get inlet fabric here, bring some with you. It does not matter which color it is. However, if you want to have good shirts, get a piece of blue and white twill made,[151] but not finer than 30 wool.[152] The warp should be 4 by 4. Furthermore, bed sheets made from linen are very comfortable here since they feel cool in the summer. You can bring some of them as well because ours here are pretty

much used up. Bring your table cloths, as you will have good use for them here. If you can, buy pewter dishes such as plates, bowls, and jars, since porcelain is rather expensive here. That is about all I would advise you to bring with you. I remain,
Your loving brother
Christian Friedrich Bergmann

[9.1]
Guadelupe, 18 May [18]67

Dear Brother:
In the meantime, I have received an order. When you come to America, do not exchange your money because a friend of mine will go to Germany and wants to exchange the money with you. You will get a better rate with him, and it will all be American money and in gold. That way you lose the least. It is Hermann Woltersdorf from Löbau. He is the son-in-law of Richter Senior whom you already know. Actually, there is also something else I would like you to bring over, namely a strong two-tiered winch and a screw for braking to be used underneath the wagon because here one cannot use it in the rear of one.

So far my prospects for a strong crop have been good. I only have corn and a few acres of cotton this year, but I am quite content that I did not plant any wheat and rye because last fall we had such an infestation of grasshoppers that hardly any wheat or rye has survived. We were lucky that in mid March we had a quite strong Norther that killed the young grasshoppers. Please also bring a screw for Friedrich and Karl each, of course, with everything necessary for mounting. Woltersdorf will not leave until June 1, so there will be enough time for you to exchange money with him. Woltersdorf stayed a whole year with me during the war time, and Richter Senior stayed almost half of his time. You will hear much more about this from Woltersdorf.
Many greetings to our father, siblings, and brother-in-law,
Your brother
Christian Friedrich Bergmann

[9.2]
Guadalupe, 16 May 1867 [Letter written by Bergmann's second oldest son Karl, born 1845]

Dear Grandfather and Godfather:
You might have wondered why I did not write to you as well. You probably think I have forgotten you, but this is not at all the case! For at the time the letter was mailed, I had not been at home for four weeks. However, since father is writing again right now, I cannot let the opportunity pass to also write you a few lines. As a matter of fact, I will get married on the 26th of this month to a girl named Emma Schladör. This year, together with her foster father who is a German by the name of Herrter, I planted 25 to 30 acres of corn (*Mais*), 3 acres of wheat, and several acres of sugar cane. I will get to keep half of the harvest. We had ten acres of wheat and two acres of rye, but the grasshoppers ate all but the three acres of wheat that are left. They would have eaten those, too, if we had not gotten a very cold Norther on March 13, which caused all of them to freeze to death. That was luck, otherwise we would not have harvested any wheat at all in our entire region. This way we could at least save the seed. Of course, the Norther caused a lot of damage to the trees because many had already begun to bloom, such as the mulberry and the peach trees. Supposedly, this was the coldest day on record in the United States.

Dear grandfather, I received the picture you had sent me and I am very happy that you are still in such good shape. I have also received the farmer's song which would be uplifting to any farmer. Thank you very much for sending it. We also wanted to send you our picture, but we have not had an opportunity yet to have one taken since the journey to San Antonio is pretty far and thus cumbersome. However, I hope that we can do it soon.

I will have to finish my letter now. Please say hello to everyone for me, to uncles, aunts, cousins, and also to my godfather C.F. Bergmann on the Haine. Tell Dammbitterlich that although we have quills and ink, we do not have much talent to use them. Thus I hope that you will forgive me but this was the reason why I always shied away from writing to you. I hope this letter will find you in good health.
Greetings from your loving grandson
Karl Heinrich Bergmann

[10.1]
Guadalupe, 30 July [18]71

Dear Sister-in-Law:
I received your letter on the first day of Pentecost,[153] but I had very much looked forward to receiving a letter from my brother since he had written me in his last letter that I was going to get a very long one. Instead it was his death certificate. We all were deeply saddened about this news. It seemed to me as if it was impossible to be true; but then, some day it will be my turn as well. I had planned to write right away, but a lot of work had piled up, and it also always takes me quite a bit of time until I finally write a letter. As a result, I had postponed writing to Sunday but then I always had visitors who kept me from doing it so that I ended up continuing to put it off.

I cannot really write you much news. This year's harvest is poor since we have not had one drop of rain since the middle of May. The last rain was a torrential storm that did not reach my land. Yet, where it hit it caused the farmers a great deal of damage. My wheat harvest was bad this year and yet, others had even less to thresh than I. My corn crop has worked out better for me than for many others because in many fields, the cornstalks have already been cut off to be used as feed and I have not needed to do that yet. I will have more than half a harvest since I planted my seeds only after the rain and so it grew when the soil could not yet all dry out.

Dear sister-in-law, I would be delighted if you could keep the farm because I would not like to hear that a stranger was going to live there. Christoph will get married around the time of the fair in Ebersbach.[154] By the way, we have a great amount of work to do because we cut shingles, and I also cut for other people when they bring me wood. I get one dollar for a thousand, so I always make a good profit. Friedrich, Christoph, and I[155] can cut 10 thousand of them in a day. However, we do not manage doing so every day because we have to keep the fire going all night long and right now in the summer, it is also too warm to stand close to such hot wood every single day. The majority of people who come to have shingles cut also want to have them readied for the market, so I get one dollar and a half for one thousand. Thus we switch between cutting and packing them which makes the work more bearable for us. I cannot write you any further news. While in recent times the Indians have been quite bad and have stolen many horses,[156] I have not experienced any loss because of them yet. However, lately the Government has taken steps that point with certainty to an outbreak of an Indian War.[157] So

the settlers at the borders are in good spirits again since in the last 3 years the Indians stole many horses and also killed people.
Greetings to your children from their uncle and your brother-in-law
C.F. Bergmann

[10.2]
[Letter written by Bergmann's wife Luise]

Dear Sister-in-Law:
I have to write you a few lines because I, too, would like to know something about my relatives and friends as they are also your relatives. When you write again, write me what your parents, siblings, and the siblings of your father are doing because I have not heard anything from Gabriel Freuden yet. I think she has completely forgotten me. You know I have mentioned her and also the others several times in my letters.

I welcomed the news of my brother's death because I could not hope that he would improve. He had fallen too deep and had become a burden to the community, which caused my father great worries since he foresaw that things would get to that point. Say hello to everyone for me, also to Heinrich, Randig and his wife, K. Dammbitterlich and family, and sister Eleonore and the church organist Michael. I am still in good health, and hope that this letter will find you in good health, too. Write us again as soon as possible.
I remain,
Your sister-in-law in the big, wide world
J.Ch. Luise Bergmann

[11.1]
[no date; written probably between 1870 and 1874][158]

Dear Father, Brothers, Sisters, etc.:
You say it has been a long time since I last wrote you, but this is not the case. I had answered your letter before the last one and told you that you should pay Schmit the interest and that I would send the money as soon as possible. I also wanted to answer the letter from California right away, but I could not get the money together; that is the reason why the whole thing took one month longer than I wanted. It really worried me greatly that I was unable to pay, but I was restricted by all sorts of losses, so I just could not do it. For the wagons I had to pay 123 dollars in fees for transportation and customs in Indianola, and then I had additional expenses for the smithy. All of this would have been okay if

the wagon made by the new wagon-maker had not broken the first time it carried a load. We were only 3 miles away from home when two wheels broke, one in the front and one in the rear. All that foolery cost me 46 dollars because I could only use the two hubs in the back; the others were too small. The chocks were too weak. Last spring, the other wagon, the one I had given to Karl, also went to pieces. Karl also is married, but he lives 30 miles further north in Comfort.[159]

Now I also want to tell you that I lost 27 trek oxen in the past 4 years, partially because they were sick, partially because they were stolen or ran away, some were killed in an accident. Among them were 7 animals for which I had paid 25 dollars each and this, of course, has set me back somewhat. Last fall it was a year since Reimann had come to me as you know.[160] He could not find work right away, so I kept him here with me until he would find work. He then went to a neighbor where he had found work and was there for 3½ months. Then he was fired because he was too rude. He then came to me again and talked me into building a shingle machine together and I agreed to it as a joint undertaking. Thus we worked on building it for 3½ months until the machine was finished. I put up the money for it. During that time, he had free food and free laundry at my house. Finally, the machine was finished, and we began to cut shingles. We had cut sixteen thousand shingles and wanted to make packages. But then things turned out differently because it began to rain, and it rained continuously for 4 days, and our shingles were swept away. That was another loss of 110 dollars, but the machine was also gone and Reimann, too. And now, Bergmann, see how you can get your money back.[161]

During that rain, the Guadalupe River rose 53 feet! I had water in the kitchen. I also had a smoke house made from stone that held up right until the last 3 hours, but then a cypress tree floated by and leveled it. I also had a lot of damage in the field and lost more than two thousand fence posts, and all the good soil of 6 acres was washed away in addition to the corn that grew on them. But the other things also suffered because the water level remained high for several days, so that only half of the harvest survived. But I comforted myself with the fact that it was not my fault and that others fared as badly. Everyone who lived on the larger or smaller rivers experienced more or less damage. All the mills were swept away; in short, all gears are either completely ruined or heavily damaged. But I forgot to mention how much in cash I had put up for the shingle machine. It was 94 dollars that I first had intended to give to Schmit before I decided to build the machine. I would not say a word if Reimann had told me that he did not want to have anything to do with the machine anymore, but he left me and went to Boerne to work in the mill where he gave me 5 dollars. From there he went to New Braunfels and has not returned to me. So I have to deal with this setback as well. He made quite a bit of money while he worked in Boerne and New Braunfels, yet I do not know where it goes. He still had a saddle at my place but he secretly sold it so that I do not even have that one left [to sell].

I have to come back to the high waters again. They caused a lot of damage to the trees along the Guadalupe River. Many shingle trees are broken or were swept away. These were trees with a diameter of 3 ½ feet. Others broke right in the middle. I had a cottonwood tree that was almost 4 feet in diameter. It is also gone because whatever the water did not carry away, the driftwood did, so I cannot even tell you the extent of the damage.

Friedrich and I[162] had been in the cedar forest shortly before the high waters, cutting wood for his house. Usually one does not take the tools home. However, by the time we could cross the Guadalupe River again and enter the forest, everything was gone, because nine weeks had passed before we could cross. Luckily, all cows were on this side. Otherwise we would not even have had butter or milk. It was a different matter regarding the trek oxen. Not one of them was on this side and that made work particularly hard. The fence was open and not one ox was in there. After the rain, there was a sweltering heat and more work for me to do than I liked. On top of it, I was sickly and could not get any worker because everyone was busy taking care of their own problems. Thus we three, Friedrich, Christoph and I,[163] collected brushwood to build a fence, but in spite of this, cattle got into the fields because in the first days, the water level of the Guadalupe River fell so fast that every night cattle came into the fields. I had a lot of damage from the cattle and, therefore, have only half a harvest. This year, I had again planted 6 acres of rye and wheat.

I think we can soon talk more positively about the political conditions here. We defeated the Democrats in the elections. The new Governor is a good Union man, and the members of Congress are also radical, as are those in the Senate. All the ones I know are capable men, so we hope that the State of Texas will also be accepted into the Union again.[164]

At Pentecost, it will have been two years since I sent Christoph to Indianola. I gave him 28 dollars, he was gone for 8 weeks, returned home sick, brought back 53 dollars, and had lost a yoke of oxen. He was sick for 4 months and, of course, no money was earned during that time. The reason for this is that I myself can no longer drive because, obviously, it is not like in Germany, where one keeps draft animals in stables overnight. I can still drive when there is someone who helps me search for the oxen in the mornings, because since I have been able to hear with only one ear, I cannot do it by myself anymore.

I can turn from side-to-side as much as I want; I still hear the bells from the side to which I turn. I just want to give you an example: I want to look for the horses, I hear the bell all the time but keep getting further away and end up feeling forced to return home after having wandered around for hours. Home again, I take my wife along (of course, the horses may not have stayed in the same location), well, I could tell you more stories about similar adventurous experiences.

Dear brother, you wrote me regarding a gold dollar. I am sending two, one for you and one for my promised godchild. In the past, I had put in 5 of them but, unfortunately, the letters have never arrived. This time I will send the letter registered mail to make sure it gets there. I would also have liked to send a photograph of myself and my wife but it has not yet fit into our schedule that we both could go to San Antonio. We were there once, but the photographer was not there, and we have left it at this.

There is something else I would like to mention. I have been plagued by a tapeworm for several years and just cannot get rid of it. I had three treatments but all of them were unsuccessful. As soon as I have time, I want to go to San Antonio to finally get rid of it for good. I always tried to avoid paying the 10 dollars the treatment will cost, and it also will take one week of my time, but I have to take care of it now. Besides that, we are all in good health. The three boys are a little overworked. Friedrich is almost as tall as I am, Karl is two inches taller than I, and Christoph is half an inch shorter than Karl, but I believe that he will end up just as tall. I want to end my lamenting. I think I am out of the woods now.

Your loving son, brother, and brother-in-law
Christian Friedrich Bergmann

Complete address:
Sender: C. F. Bergmann
P.O. Spring Branch[165]
Comal Co.
Texas

If you send a letter, mail it to Comal County, P.O. Spring Branch. It is a new post office that is not far from me.

[11.2]
[Letter written by Bergmann's wife Luise]
To Christiane Elisabeth Bergmann, Widow
6 December [no year, probably between 1870 and 1874]

Dear Sister-in-Law:
We received your letter in which you wrote about family news and have answered it, too. Furthermore, Christoph thanked you for the gift. At the time, we mailed two letters; one was addressed to you, the other one to August.

You promised that when Karl came back home for vacation he would write to us again, and so we have waited and waited, but in vain. Should he come home for a visit soon, be so kind and write us a few lines so that we may also hear what is going on. I have nothing else to write, as my husband wrote about everything else.
Many greetings to all the relatives and acquaintances
Joh. Chr. Luise Bergmann

[12.1]
Guadalupe, 4 January 1879

Dear Sister-in-Law:
I made you wait for the letter for a long time because I always waited for the photograph, but it has not yet arrived. Furthermore, I am not a big letter writer, since it always takes a long time until I finally pick up the quill to write a letter. Often I do not have the time, often I just do not feel like writing. I have to cultivate 20 acres of land and do this all by myself. Therefore, I need to be on the go constantly. The good thing is that my wife and I are both in rather good health.

Dear sister-in-law, you wrote me what the food prices in your area are, so I also want to write you what things cost here. One bushel of wheat of 60 pounds costs 60–70 cents, winter wheat 1 dollar, rye 50 cents, oat 25–30 cents, corn 30–35 cents. Beef costs 5–6 cents. Butter costs 10–15 cents in the summer, right now 30–40 cents. Eggs cost 10–15 cents in the summer, now 20–30 cents. Pork butchered costs 3–9½ cents per pound at the market, a three-year-old ox ten 10–11 dollars, a four-year-old ox 12–14 dollars. One bushel of potatoes costs 1 dollar and 30 cents, 5 gallons of beer 2 dollars and 25 cents, one gallon of brandy 1.50–2 dollars. 5 pounds of coffee, 9–10 pounds of sugar, and 8 pounds of sugar loaf each cost 1 dollar. A farmhand makes 12–15 dollars per month. A reaper gets 1 dollar per acre for the labor and one dollar for the machine each. Fabric for clothing is also cheap. 1 and ½ cubits[166] is one American yard; 16–24 yards cost 1 dollar. Fabric for pants[167] costs 15–25 cents, shoes 1½ dollars, women's shoes 75 cents to 1 dollar, at times 1½ dollars, and boots 3–4 dollars.

This year, my crop was not good. I had planted 5 acres of wheat and threshed only 18½ bushels. The reason for this was the weather conditions. The oats[168] were good, but the corn was also only a little more than half a crop. The cotton harvest was not the best either. The dry weather had hurt it quite a bit, and it is at a very low price right now. Before Christmas, it sold at 5–6 cents per pound in San Antonio; but I do not want to sell it now because prices are already beginning to go back up.

As for the Yellow Fever that raged so terribly in New Orleans and other, partially big cities, the mortality rate was higher than it had ever been during earlier Yellow Fever outbreaks. It is hair-raising to read the reports in the newspapers. The exports and imports suffered greatly because of the quarantine.

The conditions here in Texas are disheartening because, no matter which newspaper you pick up, it is full of theft and murder. Horses and oxen are often stolen at night. It also frequently happens that one's revenues from the market are stolen while one is on the road home, if one is lucky and does not get killed for them. However, this is not the case where we live or in the mountains. The mail transports and trains get stopped and raided, and this has happened rather often in recent times. The thefts of horses and oxen are usually by Mexicans. The ones who stop the mail transports and trains are elegant men of some status.

We have a cold winter this year. Although it is not horribly cold, the Norther (north wind) is constantly blowing, and it is dreary and also humid.[169] As a result, the soil in the fields is so dry that a good amount, or rather a great amount of wheat could not be planted. And now that the time approaches to put out the summer wheat, we have all that Norther. The cattle suffer, too. Friedrich and I feed the ones that return home corn, cottonseed, and oat-straws. Last spring, I bought a calf for 15 dollars (an ox-calf).

Dear sister-in-law, you complained that I had not written in a long time, but I did answer every letter I received. Samuel wrote me and asked for some seed which I sent him, and I had asked him to let me know what results he got, but I have not heard back from him. The same thing happened with Gustav Dreβler to whom I also wrote. Maybe the letters got lost, or you just have not heard about them. Friedrich bought a farm that borders ours for 500 dollars. He now has 4 children, 3 boys and 1 girl. They are all healthy. Christoph also has 4 children, 3 boys and 1 girl. Karl does not have any children. Christoph and his family are also healthy, but Karl's wife is constantly a little sick, whereas he is in good health.

Earlier in my letter, I described the conditions here in Texas and while they are indeed not particularly inviting, I would not want to be back in Germany again. We are not affected by all these things that I mentioned above. Hypocrisy and worries about having enough food do not weigh us down, in spite of the bad harvest we had. Last spring, I sold livestock for 180 dollars and this spring, I will be able to sell quite a few again. I slaughtered 4 big pigs this year and also a cow and made quite a lot of Bratwurst and other sausage.[170] Several times in the past, I have said to my wife, 'if only people in Germany were given to eat what we do not want.' Just to give you an example, the head and the feet of an ox are thrown to dogs, the other innards to the pigs, and quite often, one does not even remove the suet from the intestines. If you ever inspected our cows, you would probably say that they are pretty debauched creatures. Every Sunday, we have cake, and we are not the only ones. Everywhere one goes, there is cake. Boiled meat does not taste good, it has to be roasted. Only when one wants to make soup is meat cooked until it is soft.
I remain,
Your brother-in-law
C.F. Bergmann and L. Bergmann
Many greetings to my sisters and brothers-in-law

[12.2]
[Letter by Bergmann's wife Luise]

Dear Sister-in-Law:
I also want to write you a few lines so that you will believe that I am still alive. I wish all of you good luck for the New Year from far away! My husband has accurately described the negative and the more positive aspects here. If anyone feels like coming over, he should go right ahead and do it. Texas is still big, and decent people will get help everywhere.

At first, I was very homesick, but now I would not trade with you anymore, in any respect. That is true for all people who come here. Even those who went back to Germany soon come back. Right now, there are also many emigrants who come to Texas from the northern states because there is still so much land on which one can settle. Friedrich bought his farm for 500 dollars; it is 320 acres.[171] The buildings and the fence on the land were in poor condition, but it was cheap. In our area on this side of the Guadalupe River there are rather many Germans, and the Americans are quick to sell. They do not quite like to live amongst Germans, and the Germans feel similarly. We have railroads in Texas as well, one from Galveston to San Antonio. It takes one day to get from Galveston to San Antonio. They also built a railroad from Houston to San Antonio that goes through Austin.[172]

I want to finish this letter and wish that it will find all of you in good health. Please say hello for me to all my relatives, acquaintances, and friends. Tell them that I am doing well.
Farewell. I remain,
Your sister-in-law Joh. Chr. Luise Bergmann

[12.3]

Dear Nephew:
I saw from your letter that you also want to become a teacher and, in my eyes, you are right, and I wish you good luck with it. What you wrote about coming to America seemed like a dream to me when I read it. Well then, come on over, from Galveston to San Antonio it is only a day trip, and it is not more from San Antonio to our place. But familiarize yourself a little bit with the English language, or else you will be considered a greenhorn, just like everyone else. Your father also had wanted to visit me here in America, but he died too early. I am sure you will write me to ask what you should bring, if there is still enough time to do that at all. Just remember that I will celebrate my sixty-second birthday next April and that my wife celebrated her sixty-first birthday right before Christmas. It will feel like a long time to me until I will see you, but I do hope that I will still be around for it.
Your uncle
C.F. Bergmann
Guadalupe, Kendall County

[12.4]

Dear Nephew and Godson:
As you wrote me, your situation has changed considerably, and I am also glad that you are doing well. I am sure you would like to know a lot about Texas, but I already wrote everything in the letter to your mother.

I would have liked to see you in Texas, but now I have to give up that hope. A teacher in San Antonio makes 57–100 dollars per month and in the country 20, 30, and 40 dollars, depending on how many children he teaches.

A teacher here may not teach as many students as a teacher in Germany, but here they also have to stay at school for the whole day. Those who live far away from the school have to bring their lunch to school.

There are also singing societies here in every larger settlement and, as experts on this subject say, they are well accomplished. Every year, they have a big song festival. The last one, in which song societies from Galveston, Houston, Austin, Fredericksburg, Comfort, Boerne, and other places participated, was in San Antonio and lasted for 3 days.[173] There were also two bands that took turns in making music. There is no other news I could write you, but I am sure you will read my letter to your mother in which I described things the best I could.

Happy New Year,
Your uncle and godfather
C.F. Bergmann
Guadalupe
Kendall County

[13.]
Guadalupe, 28 May 1880

Dear Nephew:
I had you wait for a letter for a long time. I meant to write a lot to you, but not much has happened here. Furthermore, I am more skilled in using a plow and a hoe than a quill because it takes me a long time to finish a letter, especially with a trembling hand, as you can see from this letter. After all, I will turn sixty-three next April and am not used to writing very often. I received your letter and the photographs and thank you for them. It was a great joy to see the house of my parents again, but I was also happy to see Organist Ostritzer on the picture. The news of your mother's death came as quite a surprise to us. As far as my sister is concerned, I wished she would get a good beating.[174]

We had nearly no harvest this year here in West Texas. From Christmas to September, it rained only twice, therefore we could not plant any wheat, and those who did plant wheat did not see the seed germinate. Although I harvested as much as I need for my own farm, I have nothing left to sell. However, you will not have to imagine that we are in need. I butchered four pigs and a cow, and I wished that the poor people would have what we throw to the dogs. For here one does not use the head, heart, and lung, nor the feet and the chitterlings of cattle. In short, we are not starving, and I can also enjoy a glass of wine every day. This year, I made two hundred quarts (1 Quart=1.136 liters)[175] of wine. Obviously, it is not a wine from the Rhine River or a Burgundy, but it is drinkable.[176]

We received sad news from Germany already last summer, namely about the widespread rain. Now I just read in the newspapers from January 16, that all of Europe has suffered severe damage from the ice drift. Here in America, there has also been much damage due to the elements. In many states there were tornadoes that caused enormous damage. Already back in the fall, we read about all the snow and cold weather in Germany, and now the political turmoil and hunger in Silesia and in parts of Saxony.

We too had quite cold weather at Christmas. In many places, the temperature had dropped to 17 degrees below freezing, and quite a few people froze to death who partially had bought themselves a [?][177] for Christmas, and also some who were traveling. I had planted 4 acres of winter oats, but the frost destroyed everything. By the way, the farmers put out a lot of seed, but they were able to do this only after Christmas because it was always too dry to plow.

Wait, what is this? A postcard. You are right in requesting an answer in this way. I was just at this point in my letter when I received it. We can expect to have a good harvest, so we will be compensated for last year's bad one, and we are beginning to reap the grain. On April 11, Karl's wife died. She had been bedridden for 7 months.

In America, forest fires and wildfires are the biggest news. 50,000 acres of forest burned in the state as well as many nearby villages, and a number of people died in the fire. Do not hold it against me that I did not write a letter to you. I really would not have known what to write, and I do not want to make up things. In 14 days, the big aloe tree will bloom, and I wished you could see it. The stem has reached a height of 14 feet and has started to show buds; it has a diameter of 7 inches.

You also wanted to know whether I often get together with my children. Friedrich I see every day, since he lives only 600 steps away. I also meet Christoph rather often, since he lives only 10 miles away. Of course, it is different with Karl who is 33 miles away in Comfort. He was here at Pentecost, for the first time since Christmas two years ago. Now that his wife is dead, he has promised to come over more often. Before, it used to be the opposite, and we had to go to him.

You want to know how long it takes us to get to San Antonio. It takes me one day to get there and one day to come back. Our letters and newspapers we have to pick up ourselves. Lately, things have been more convenient since we now have a post office very nearby. So, when you write again, the address is P.O. Guadalupe, Kendall Co., Texas

I decided from now on I will write down all thoughts that I would like to share in my next letter as they come to my mind. My memory is getting short. I hope that neither one of you will mind that I did not write individual letters to each of you, especially since I would have written exactly the same news in both of them. I am also sure that neither one of you will withhold the letter from the other person.

On May 28, we had another flood. Again, the Guadalupe River had risen 16 feet within 6 hours from midnight of the 27th until midday of the 28th. However, it did not cause any damage. Of course, the heavy rain took much of the harvest into the Guadalupe River.

I want to finish my letter now and hope that it will find you in good health.
Greetings to both of you,
Your uncle
Ch.F. Bergmann
L. Bergmann

[14.]
Guadalupe, 8 December 1881
Kendall County

Dear Nephews:
I already had bought a postcard to ask you for a response when I received your card, because I had written a long letter to you without getting an answer from you. Well, I do not have much to write you today either. The two of us are, thank God, fairly healthy, and I also had a decent harvest. Indeed, there were not many who did as well as I did with my corn. Many people did not work hard enough, others could no longer work in the corn fields due to the drought. It was literally unbearably hot and dry.

Last Sunday, Karl remarried. She is a hard-working farmer's girl from Boerne.[178] On that occasion, there was a lot of singing and eating. Karl is also a member of the singing society in Comfort.[179] I attended the last song festival in Boerne as well.[180] There were 7 singing societies there, and the festival lasted for 3 days and 3 nights. You can probably imagine all the things that were going on there. Music bands and choirs took turns and the last evening, there was a torchlight parade followed by more social gathering. The nicest thing about these festivals is the fact that there are no [rich][181] or poor, everyone is the same.[182]

Dear nephew, you would like to know whether one dances here in the same way as in Germany. Yes, we have the same dances that you have, and some others, except whoever wants to dance has to pay an admission fee of 50 cents to 1 dollar, depending on the number of musicians. But then, the music starts already at three o'clock in the afternoon and goes on until the next morning. We were at the last May festival as well, and my wife and I amused ourselves quite a bit because old people, too, have a kind of need to let off steam once in a while. I am certainly not the one who would miss such an opportunity. I am also in a better position now than 12–15 years ago to afford such entertainment, because there is a big difference between getting interest and paying interest. Friedrich, my oldest son, is also very content with his life situation. He has almost all the farm equipment a farmer needs. Last summer, he bought a harvester, and next summer he will buy a thresher. Then he will have everything that one needs on a well-equipped farm. He also succeeded in digging a well. Sure, it cost a lot of money because he had to blast through 38 feet of rock, but now he is happy that he has water. Christoph, too, lives in rather good conditions. Sure enough, he is still quite a young farmer, but he also works hard and I think he will be better off over the years. Actually, he also makes a good amount of money in his smithy. He made himself a bellows and bought himself an old smithy. As for Karl, we all hope that a new era will start for him in his new marriage. He certainly did not get very far with his first wife since the doctor and pharmacy bills always ate up the biggest part of his profits. For that reason all of us were happy that he married a hard-working farmer's girl.

I could not make wine this summer, since it froze with the last frost of the spring, so we can only think of it. Now it is about time again to butcher pigs, and this year, I have only 3 to butcher, but they are rather big, so that I can sell two of them, yet I will make quite a bit of sausage again. The prices for cattle are good right now. A one-year-old calf costs 40 marks,[183] a two-year-old costs 60 marks. Large oxen for slaughter are in great demand and hard to find. A fat cow costs 72 marks, an ox for slaughter costs 100 marks, a cow with calf 72–100 marks. By next spring, I want to sell 10 cows with calves. That will be my beer money, because I really like to drink beer.

I still need to tell you about a recent adventure. After the wedding, I went to Comfort with my wife. Of course, we were in an excellent mood and had taken the necessary drinks with us as there are no inns along the way. When we finally arrived at the area where we wanted to have our drinks, we stopped and opened them. After we had finished, we continued our trip, but destiny, that is, the rear wheel, had a different plan and wanted to be in Comfort before

us. It went straight down the hill with a piece of a shank, and we had to go on riding on three wheels. Once again, there was a big Hurray for Texas. We did not have an axe in the wagon, but it did not take long for us to get one and, an hour later, we continued our ride. Another hour later, the long main beam broke, which caused us to stop again while having a great laugh. Finally, once we arrived in Comfort, everyone laughed and was sorry for me because we had made loops and had been half-way sledding. Now we are at home again and go about our work. I won 25 dollars at the last fair. You probably saw my name in the *Farmer* Presse,[184] and you will be able to read more about it there.
Greetings from your uncle C.F. Bergmann and L. Bergmann
and all 3 sons

[15.]
Guadalupe, 12 March 1885

Dear Nephew:
It has been a year since I received your postcard in which you wrote that you were going to write a long letter. I have waited for it in vain and think it is about time for you to write one. Before Christmas, it was 3 years since I mailed a letter to you, but I have not received a response yet.

We had an extremely cold winter. So far I have lost 14 cattle and expect to lose more because there was no grass and the extremely cold weather with freezing rain prevented the cattle from being able to feed on anything. Since there was nothing on which they could feed, we had to feed the cattle throughout the whole winter. But I am not the only one, I still am lucky. Someone who lives 12 miles away from me lost 500 out of his 1,400 cattle, and others suffered a similar fate. There is not a single person who did not lose anything.

Last year, Friedrich bought a threshing machine for 656 dollars. He has already threshed close to 8,000 bushels and intends to build a new house this summer. Karl and his family are well, too. He has two boys. Christoph and his family are also healthy, as are my wife and I.

The last winter caused quite a bit of damage here in the United States, on one hand due to the snow and cold, on the other hand due to earthquakes. Now everyone is worried about the time the snow will melt. If there is a lot of rain, the big rivers will likely rise and cause much damage again, because we had enormous amounts of snow. The last harvest was not particularly good. As a result, prices for grain are not good. The bushel corn goes for 50 cents or two marks, wheat for 60–70 cents, oats for 30–40 cents.

Other than that, we are doing fine. I can drink wine every day. Last year, I made two barrels of wine and still had one older one. One barrel holds 40–45 gallons; a gallon is four quarts, so I have plenty of good wine. Now springtime[185] is approaching, and we are beginning to work hard again. The oats are already rather green. Also, I had a well made because we got tired of carrying the water up from the Guadalupe River. It did cost a large amount of money because we had to blast through the rock,[186] but we are happy that we have a well now. It cost close to 200 dollars.
Now I am anxious to see whether one of you two will have time to send me a few lines in return.
Greetings[187] from your uncle and aunt
C.F. Bergmann and wife

I had intended to take you to task but then I thought it would not help anyway.[188]
If you get a chance, tell my sisters that we all still are in good health.

[16.]
[Handwritten obituary for Christian Friedrich Bergmann by the Germania Farmer Association; 2 December 1888][189]

C.F. Bergmann

died on the 24th of the month[190] around eight o'clock in the evening as a result of a painful gastric disease due to which he was bedridden for the past two months.

There are increasingly fewer old pioneers who bore farewell to their German homeland—for the most part, in hopes of giving their children a more humane life than they would have had back in the old country, who came to settle here in this area which, at the time, was still desert and barren and a stomping ground for rapacious and murderous Indians. While enduring indescribable hardship and suffering need, they turned this land into blossoming homesteads.

The deceased was privileged to reap great rewards for his efforts, in which his surviving wife enthusiastically supported him.

He was honest and upright, truthful in all his actions, fearless and absolute when it came to fighting for justice, which he defended against each and everyone. He was German through and through in his thoughts, as well as in his words and deeds, while conscientiously carrying out his duties as a citizen of the Union. He also succeeded in instilling these traits in his three sons who, together with their wives and grandchildren, form a large family circle at the center of which he was the patriarch.

Bergmann was born in the year of 1817 in Ebersbach, Kingdom of Saxony. He came to Texas in 1854 and initially stayed in San Antonio. Soon, however, he settled along the Guadalupe River, where he built himself a comfortable home. There he died and was buried. Unfortunately, the bad weather caused the cancellation of the magnificent funeral feast that was planned.

Decree of Condolences

Due to the death of our dear fellow citizen, C.F. Bergmann, on November 24 of this year, the Germania Farmer Association determines that through his death we have lost a valued member, a true friend, and an open-minded, upright and honorable person. We deeply mourn our loss and will faithfully preserve his memory. We herewith express our sincere condolences to his dear wife and family members. Furthermore, we decide that these resolutions be entered into the minutes of the Germania Farmer Association and that a copy of the document be given to his mourning widow and be printed in the *Freie Presse für Texas*,[191] the *Neu Braunfelser Zeitung*[192] and the *West-Texas-Pionier*.

Anhalt, 2 December 1888
F. Haffring
Hans Wehe
C.H. Hofring
Committee

Notes

[1] This was the first letter Bergmann sent upon his arrival in America. The end to this letter in the original is missing. The letter does not have a date; it was most likely written in the months following the family's arrival in San Antonio on November 18, 1854. Bergmann refers to this letter in his letter of 3 December 1856. The style of the letter indicates that Bergmann seems to have taken notes during his sea journey.
[2] Another German immigrant who came to Texas on the same ship. Later in this letter Bergmann mentions that Heuchling's youngest daughter had died during the sea journey. In a letter of 1 December 1859, he describes how Heuchling first seemed to do well, but that things–as it seems–then fell apart for him.
[3] Leipzig is a city located in the federal state of Saxony, Germany, about 95 miles south of Berlin. Since the middle of the 19th century, Leipzig was a hub of railway traffic in Central Europe.
[4] *Zum schwarzen Kreutz* is the name of a restaurant in Leipzig that existed from 1704 to 1863, located in Gerberstraβe 10. Known today as *Brüsseler Hof.* Cf. <http://www.leipzig-lexikon.de/index.htm>, [Accessed May 12, 2013].
[5] The harbor of Bremerhaven, about 31 miles north of the city of Bremen, was an important harbor for emigrants to North America. Between 1832 and 1974, over seven million emigrants left Europe from Bremerhaven, 90 percent of which were bound for the United States of America. Cf. "European Emigration. The Historical Records," <http://www.european-emigration.com/de/warten.html>, [Accessed May 9, 2013].
[6] Probably emigrants were a target for thieves.
[7] The distance from Leipzig to Braunschweig is about 127 miles. Braunschweig (in English: Brunswick) is a city located in the state of Lower Saxony. Since 1498, the Braunschweig Fair was a biannual trade fair.
[8] The distance from Braunschweig to Bremen is about 239 miles. Bremen is a hanseatic city in northwestern Germany.
[9] This was a paddle steamer, a steamship powered by a steam engine using paddle wheels to propel it through the water. There were two paddle steamers of this name. The *Roland I* was built in 1839, the *Roland II* in 1847/48. Most likely the family used *Roland II.* Cf. Hans Szymanksi. *Die Dampfschiffahrt in Niedersachsen und in den angrenzenden Gebieten von 1817 bis 1867* (Bremen: Europäischer Hochschulverlag, 2011), 24 and 31.
[10] They travelled on the Weser River from Bremen about 31 miles north to Bremerhaven, where the Weser River empties into the North Sea. The Emigration House was built in the years 1849–1850 and had a size of 31,215 sq. ft. It could house 1,500 to 2,000 people. Cf. Anja Benscheidt, Alfred Kube: *Bremerhaven und Umgebung 1827–1927. Geschichte im Morgenstern-Museum, Band 1* (Bremerhaven: Nordwestdeutsche Verlagsgesellschaft, 1993), 49. For engravings of the Emigration House see the German Emigrants Database. Historisches Museum Bremerhaven: <http://www.deutsche-auswanderer-datenbank.de/index.php?id=184; <http://www.deutsche-auswanderer-datenbank.de/index.php?id=132>, [Accessed May 8, 2013].
[11] In Bremen-Vegesack, northwest of Bremen, the Lesum River, a right tributary to the Weser River, flows into the Weser River.
[12] The ship was a brig, a sailing vessel that has two square-rigged masts. The brig *Minna* was built in 1851–52 in the shipyard of St. Magnus near Bremen, Germany, on the order of Captain Hinrich Wessels in Flethe. Its port of registry was Blumenthal Flethe. The brig had a cargo carrying capacity of 270 register tons. One register ton is equivalent to 100 cubic feet of cargo

space. Its maiden voyage which began on 3 March 1852, was from Bremerhaven to New York City. In 1863, the *Minna* suffered shipwreck while carrying a cargo of wood from Quebec, Canada to Europe. Cf. Friedrich Spengemann. *Die Schiffe der hannoverschen Weserflotte der ehemaligen hannoverschen, später preußischen Ämter Blumenthal, Lehe mit Hafen Geestemünde, Lesum und Dorum und der früheren adelichen Gerichte Lesum und Schönebeck [von der Frühzeit der Seeschiffahrt bis 1901]* (Bremen: Vegesack, J.F. Rohr, 1936), 36–37.

[13] The actual name was *Magnet*. It was a tugboat, since 1853 used to tow large vessels in the Bremerhaven area to the mouth of the Weser River. Cf. Hans Szymanksi. *Die Dampfschiffahrt in Niedersachsen und in den angrenzenden Gebieten von 1817 bis 1867*, 391.

[14] The original letter states "Ostsee" (Baltic Sea), but it was actually the North Sea.

[15] Christoph Wilhelm was Bergmann's youngest son who was five years old at the time. He was born on 7 April 1849.

[16] Karl Heinrich Bergmann's second son who was nine years old at the time. He was born on 19 September 1845. The passengers on the ship usually were crowded in the dimly-lit space between decks. Cf. W. Phil Hewitt, Terry G. Jordan et alii, *People in Texas* (Boston, MA: American Press, 1981), 7.

[17] Saxon Switzerland is a hilly area around the Elbe River valley southeast of Dresden in Saxony, Germany.

[18] On 27 March and 28 March 1854, Britain and France had declared war on Russia during the Crimean War that lasted from October 1853 to February 1856.

[19] A steamship, powered by a steam engine, using one or more propellers, also known as screws.

[20] It is not clear what fish this might have been.

[21] In German "Schweinfisch." Pigfish belong to the group of scorpaeniform fish. They don't have scales and their snouts are projecting with a small mouth.

[22] Most likely Bergmann refers to the "Nordkaper," the North Atlantic right whale that can measure up to 60 ft in length. This whale does not have teeth, but long plates hanging in a row from the upper jaw.

[23] Dysentery is an inflammatory disorder of the intestine, especially of the colon, that results in severe diarrhea, fever, and abdominal pain.

[24] The Canary Islands are a Spanish group of islands, located just off the northwest coast of mainland Africa.

[25] In German "Zahnfieber;" usually, the fever is not related to teething, but the result of a bacterial or viral infection that causes the body temperature to rise.

[26] Old measurement. One German geographical mile was about 7,420 m = 4.6 statute miles, i.e. 10 German miles are close to 47 miles.

[27] German word for the spiny dogfish (*Squalus acanthias*), a temperate, coastal shark in the Atlantic and Pacific oceans.

[28] Herons are long-legged freshwater and coastal birds.

[29] Probably a cormorant

[30] Island in the Caribbean. Cuba was a territory of Spain until the Spanish-American War ended in 1898; it gained formal independence from the U.S. in 1902. Cuba is about 1,500 statute miles north of the equator.

[31] 82°F

[32] 100°F

[33] Island in the West Indies in the Caribbean region with an area of 108 square miles.

[34] Montserrat is a Caribbean island that is British overseas territory. It is part of the chain of islands known as the Lesser Antilles in the West Indies and measures approximately 9.9 miles in length and 6.8 miles in width.

[35] Island located in the northeastern Caribbean.

[36] Colloquial term for typhoid fever. It is a bacterial disease that is transmitted by the ingestion of contaminated water or food.
[37] Santo Domingo was the capital of the independent Dominican Republic, founded in 1844.
[38] Cf. n. 25.
[39] The English translation would be 'monsters.' It is not clear to which kind of fish Bergmann refers.
[40] September 17–19, 1854: A hurricane hit Matagorda/Galveston. The main impact of the storm was around Matagorda and Lavaca Bays. The storm led to a Yellow Fever outbreak, which spread from Indianola to other nearby communities over the next several weeks. Cf. David Roth, "Texas Hurricane History, National Weather Service, Camp Springs, MD," 17. Cf. <http://www.hpc.ncep.noaa.gov/research/txhur.pdf>, [Acccessed May 15, 2013].
[41] Jamaica is situated in the Caribbean Sea, about 90 statute miles south of Cuba. It is 4,240 square miles in area.
[42] Old measurement. One geographical mile was about 7,420 m = 4.6 statute miles. Thus 40 geographical miles would be 184 statute miles. The actual length is 145 miles.
[43] Probably a reference to the Blue Mountains, the longest mountain range in Jamaica. They include the island's highest point, Blue Mountain Peak (7,402 ft).
[44] Cf. n. 25 and n. 38.
[45] The lighthouse he mentions is the Roncali Lighthouse, located at the Cabo de San Antonio (Cape San Antonio), Cuba's westernmost point, on the Guanahacabibes Peninsula in Sandino. Kobantan is probably how Bergmann understood the word 'Cabo de San Antonio.' The lighthouse was built in 1850 by José Pérez Malo and has a height of 74 ft.
[46] The City of Galveston was founded in 1839. The Port of Galveston was established as a port of customs entry on 17 October 1825. It was the major port of entry for immigrants in the 19th century, but not the only port in Texas. Records indicate that the brig *Minna* arrived in Galveston on 3 November 1854. See Ethel Hander Geue, *New Homes in a New Land. German Immigration to Texas 1847–1861* (Baltimore, MD: Clearfield Company, 2002; originally 1970), 40. Yet the ship had to move on to a different harbor due to a Yellow Fever outbreak.
[47] According to the *Annual Report of the Supervising Surgeon-General of the Marine-Hospital Service of the United States for the Fiscal Year 1895. Washington: Government Printing Office, 1896,* there were 404 deaths in Galveston for the 1854 Yellow Fever epidemic; for 1853, the previous year, it lists 534 deaths from Yellow Fever in Galveston (437–38). The report encourages "intelligent sanitary work" in the Southern seaports, namely, the providing of a thorough system of drainage and sewage, good water supply and municipal cleanliness." (428) Cf.<http://books.google.com/books?id=AI9IAQAAIAAJ&pg=PA438&dq=Yellow+Fever+Galveston+1854&hl=en&sa=X&ei=WH7bUP_mDcLC2QX0_YDgCA&ved=0CFUQ6AEwBQ#v=onepage&q=Yellow%20Fever%20Galveston%201854&f=false>, [Accessed May 9, 2013].
[48] For more information on Indianola see Brownson Malsch, "Indianola, TX," <http://www.tshaonline.org/handbook/online/articles/hvi11>. Indianola was located about twelve miles southeast of Port Lavaca in Calhoun County, TX. From its founding in 1846, Indianola served as a major port and was Texas' primary port after Galveston. Many ships that came to Galveston continued on to Indianola, 100 miles further down the Texas coast. In 1867 it had around 2,200 inhabitants. Cf. Andreas V. Reichstein, *German Pioneers on the American Frontier. The Wagners in Texas and Illinois* (Denton, TX: University of North Texas Press, 2000), 148. Two hurricanes, in 1875 and 1886, and a subsequent fire destroyed Indianola and it ceased to exist. For more information on the impact of the hurricane of 1875 on Indianola, see Andreas V. Reichstein, *German Pioneers*, 150–54.
[49] 6 November 1854

[50] The emigration ship *Neptun* heading for Galveston, TX, had left Bremerhaven on 1 September 1854, and had arrived in Galveston on 3 November 1854. Cf. Immigration Lists, <http://www.fbgtxgensoc.org/lists/lists_frame.html>, [Accessed May 8, 2013].
[51] Some records list a ship of that name as having departed from Bremen and arrived in Galveston on 20 November 1854. Cf. Ethel Hander Geue, *New Homes in a New Land. German Immigration to Texas 1847–1861*, 39. Obviously, there is a discrepancy in information that is most likely due to the fact that because of a hurricane in Galveston in 1900, many records were lost. Similarly, many ship records in Bremen had been destroyed during the Second World War.
[52] In Texas, all north winds which come up suddenly, are called 'northers.' The term is principally used for the sudden north winds of winter, but also for sudden northern wind thunderstorms in the summer. See De Witt Clinton Baker, *A Texas Scrap-Book: Made up of the History, Biography and Miscellany of Texas and Its People* (New York, Chicago, New Orleans: A.S. Barnes and Company, 1875), 367.
[53] In the original letter "*Ruth* (Road=Straβe)." This is an indication that his vocabulary was already impacted by the English language that surrounded him.
[54] According to the census in 1850, San Antonio had a population of 3,488, the census of 1860 showed a population of 8,235. By then San Antonio was the largest town in Texas. Germans greatly contributed to its growth and until after 1877, there were more German speakers than both Hispanics and Anglos. By 1880, San Antonio had a population of 20,550. Cf. T.R. Fehrenbach, "San Antonio, Texas," Handbook of Texas Online, <http://www.tshaonline.org/handbook/online/articles/hds02>, [Accessed May 8, 2013].
[55] It was common for German immigrants to receive help from German settlers who had arrived earlier in getting oriented.
[56] In the original letter "gerent (gezahlet)."
[57] The end to this letter is missing.
[58] The title page of the letter is like a social mirror of San Antonio in the middle of the 19th century, by showing not only specific places, but also various ethnic groups represented in the city. Their respective poses reveal much about the social status and the power struggle of the various social groups at the time. Bergmann had written the German words for the various representatives underneath their depictions. Clockwise they read: Indian (in German: "Indianer"), Mexican (in German: "Mexikaner," American (in German: "Amerikaner"), African American (in German: "Neger"). The latter term was not a slur against African-Americans, but a term commonly used in Germany to describe a black person. Bears and wildcats could be found in Texas in the 19th century. Cf. Theodore Gish ed., *The Diary of Hermann Seele and Seele's Sketches from Texas* (German-Austin, TX: German-Texan Heritage Society, 1995), 447 and 451. This title page with views of San Antonio can also be seen in a letter of 1856, written by Swiss immigrant Henry Baumberger. Cf. <http://drtlibrary.wordpress.com/2009/10/16/how-old-man-henry-and-his-family-are-getting-along-in-america/>.
[59] In the German text literally "I and my wife."
[60] Bundesmann seems to have been another German immigrant in Texas who came from the same region in Germany from which Bergmann came. He often was involved in helping Bergmann take care of money transactions to and from Germany.
[61] It is likely that two letters following his first letter about the sea journey never arrived.
[62] Christian Friedrich was Bergmann's oldest son. Born on 20 August 1842, he was fourteen years old at the time.
[63] Based on a later letter of 12 June 1857, written by Friedrich to his grandfather in Germany, this amount refers to a monthly income.
[64] Bergmann's kids seem to have attended a Catholic school in San Antonio; possibly it was part of the Marianist St. Mary's University.

[65] Due to some illegible words in the original letter, the content of this sentence is not altogether clear. Gambling, horse racing, and drinking were common amusements in Texas and often led to violent confrontations; often it took little provocation for shoot-outs to happen. Cf. Randolph B. Campbell, *Gone to Texas. A History of the Lone Star State* (New York and Oxford: Oxford University Press, 2003), 231.
[66] Like many immigrants in the ante-bellum period, Bergmann worked as an employee doing farm work until he had the monetary funds to start farming on his own. It was not uncommon that newly arrived settlers worked for Germans who had come to Texas earlier, in order to earn money and to gain experience. Cf. Terry G. Jordan, *German Seed on Texas Soil*, 186.
[67] Bergmann seems to engage in an activity with which he was quite familiar and that was common among German immigrants. Hay was an essential part of German agriculture where it was needed for feeding the livestock during the cold winter months. Cf. Terry G. Jordan, *German Seed on Texas Soil*, 80–81 and 141. Later Bergmann will learn that livestock in Texas grazes on the open range all year so haymaking was not essential in that respect. In a letter of 1 December 1859, he explains that he feeds his cattle only salt and that they do not expect to get anything else.
[68] Everyone in the family contributed to improving the financial situation of the family. Undoubtedly, Bergmann's wife could not be choosy, but had to accept even menial jobs so she could help the family pay their bills.
[69] The German word "Korn" usually refers to grain, but as Bergmann explains in his letter of 21 June 1857, the term "korn" is used for corn. We can assume that in this letter he also means 'corn,' since he usually refers to grain in German as "Getreide." There are indications that German immigrants early on adopted growing corn and eating corn when they came to the U.S. On other fields they grew wheat and only rarely rye; the latter they used to grow in Germany. Cf. Helbich and Kamphoefner, *News from the Land of Freedom*, 60. As Terry G. Jordan points out, corn was a new and unfamiliar crop that virtually did not exist in 19th century Germany. German immigrant farmers and southerners raised large quantities of corn crops in the 19th century. Cf. Terry G. Jordan, *German Seed on Texas Soil*, 66 and 123.
[70] In the German original "Bathaten (sweet potatoes)." At the time, German lands were completely unfamiliar with the sweet potato, but Germans in Texas adopted it rapidly. Cf. Terry G. Jordan, *German Seed on Texas Soil*, 71 and 138.
[71] The white potato was an important part of of agriculture in Germany. Many German settlers believed that the white potato did not grow well in the warm climate. Still it seems that in the late 1870s and 1880s the production of white potato expanded in the area. Cf. Terry G. Jordan, *German Seed on Texas Soil*, 136.
[72] Comanche Indians were thoroughly in control of the Hill Country. Being on the frontier, the settlers were directly in contact with the Indians. Franz Kettner makes similar observations for the vicinity of Fredericksburg, in a letter of 23 April 1855. See Ilse Wurster, Charles A. Kettner, eds., *Die Kettner Briefe: The Kettner Letters: A Firsthand Account of a German Immigrant in the Texas Hill Country 1850–1875* (Wilmington, DE: Comanche Press, 2008), 49.
[73] It was a common complaint among settlers that Indians stole horses and other animals.
[74] According to Terry G. Jordan, Germans typically planted fruit trees, common in Europe. Among them were apple, cherry and pear trees, but only peaches and some native pecans, plums, figs, and apricots yielded successful results in the long run. Cf. *German Seed on Texas Soil,* 139.
[75] In rural areas it was a tradition to give the bride a cow as a wedding gift. In German it is called "Brautkuh."
[76] Theaters were common in towns in Texas. If the town was large enough, like Houston and Galveston, it could attract professional actors. In smaller towns there were amateur theatrical groups. Cf. Randolph B. Campbell, *Gone to Texas*, 230.
[77] A specific area in Bergmann's hometown Ebersbach, Germany.

[78] In the middle of the 19th century, monetary unity in Germany did not yet exist. Common currencies were the *Thaler* and *Gulden*. From 1857 to 1873, the *Vereinsthaler* was the currency in Saxony where it consisted of 30 *Silbergroschen*. Each *Silbergroschen* was divided into 10 *Pfennig* (pennies).
[79] Probably he means 'loafers.'
[80] This is an interesting comment. At the time, there was the view that camels were suitable beasts of burden on the Texas frontier since the animals needed little water, ate almost anything, and could carry as much as six hundred pounds over long distances. As a result, in 1855 the army bought thirty-three camels in Africa which were brought to Indianola in April of 1856 and taken to Camp Verde a little northwest of San Antonio. In the following year, another forty camels were imported. They were successfully used by the army in exploring the Big Bend area southwest and on expeditions to New Mexico, Arizona and California. Nevertheless, the use of camels came to an end after the Civil War, for their smell was considered awful and they caused horses to spook. In addition, Americans, used to dealing with mules, were not accustomed to handling camels. Most camels that survived the Civil War were either bought by civilians or ended up roaming free in the west. Cf. Randolph B. Campbell, *Gone to Texas*, 199.
[81] This is a common phenomenon in immigrant families. Immigrant children often embrace the new language much more quickly than their parents and frequently help the latter to navigate through their new linguistic environment.
[82] A collection of songs.
[83] In German "Gang in die Heidelbeeren."
[84] In German "Anfall von kaltem Fieber," most likely a reference to malaria, a mosquito-borne and potentially deadly infectious disease accompanied by chills, fever and sweating that recur at regular intervals. It was common among newcomers to Texas to contract malaria during spring or summer. Cf. Randolph B. Campbell, *Gone to Texas*, 219.
[85] Before the Civil War, there were numerous doctors in Texas; yet, they were often unable to cure diseases and were rather expensive, charging three to five dollars for a house call. In 1853, the Texas Medical Association was founded, but there were still many "medical quacks" at work in antebellum Texas. Cf. Randolph B. Campbell, *Gone to Texas*, 220. The fact that the Bergmann family did not have a doctor might have had to do with the high doctor's fees.
[86] The German original states "Komalkundi," an interesting rendition of an English word the meaning of which Bergmann did not seem to quite understand yet. Comal County was created in 1846. Rudolph Leopold Biesele, *The History of the German Settlements in Texas*, 1831–1861, 129 and n. 39. The census of 1860 shows that 85 percent of the people in Comal County were German (159).
[87] In German the word is spelled "Guattlup." Already in 1850, there were more Germans than Americans in Texas who were landowners. Cf. Walter D. Kamphoefner, Wolfgang Helbich, Ulrike Sommer eds., *News from the Land of Freedom. German Immigrants Write Home*, 60.
[88] In 1857, Bergmann Senior pays $200 for 320 acres. The price per acre was less than 63 cents. In 1879, Friedrich buys the same amount of acres that border his father's property for $500, thus paying about $1.56 per acre. For land prices in the Hill Country compare also Jordan, *German Seed in Texas Soil*, 189–90. The Republic of Texas had little money, but plenty of land. In the late 1830s, they began to create funds through taxation, and land was granted to new settlers from the United States or other countries. Cf. Ethel Hander Geue, *New Homes in a New Land. German Immigration to Texas 1847–1861*, 10.
[89] A bushel is about 36.2 L dry volume.
[90] As a fairly new immigrant Bergmann has to buy corn and minds high prices. Later, he will look at them from the perspective of the farmer who wants to sell corn at a high price.
[91] At the time, Friedrich, born in August 1842, was 14 years old and Karl, born in September 1845, still 11 years old. It seems that it was a practical decision Bergmann made in having Karl

confirmed at such a young age, since confirmation usually occurred more around age 14 and also meant a coming of age.

[92] This would equal a yearly interest rate of 31.2 %. Obviously, these expensive kinds of loans were intended as short-term loans. In a letter of 8 December 1881 [14.], Bergmann will allude to this by writing that "there is a big difference between getting interest and paying interest."

[93] This is an example of German immigrants who decided to return to Germany.

[94] In German, "I and my brother."

[95] Today this area is known as the Hill Country.

[96] Originally, fences were built of cedar or oak wood. In 1859, cedar fences were still the preferred material for fence-making, but at that time Germans began to build stone fences that became common in decades to come. They used the limestone and sandstone which were widespread in the area. Hiring people to build those stone fences was expensive so entire German families sometimes collaborated on building them. Jordan mentions that by 1860, Germans had erected an estimated length of several hundred miles of stone fences in Comal County alone. Cf. Terry G. Jordan, *German Seed on Texas Soil,* 164–65.

[97] In the woodlands of Germany, the right to hunt was confined to the privileged. Cf. Terry G. Jordan, *German Seed in Texas Soil*, 34. This suggests that Bergmann's free access to hunt wild game was considered by his family as a sign of great freedom.

[98] Literally "which do not understand a joke." Probably they were watchdogs.

[99] A Dresden bushel was between 103.8 and 127.7 L dry volume. A bushel is about 36.2 L dry volume.

[100] Bergmann's mention of cotton belongs to the few references regarding cotton in ante-bellum times. As Terry G. Jordan mentions, cotton came later to the German settlements in the Hill Country, experienced an increase in production by 1870, and then gained great popularity in the 1870s. Near Comfort in Kendall Country, cotton farming began on a commercial basis in 1875. In Comal County the number of gins rose from one in the mid-1850s to twenty by 1885. Cf. Terry G. Jordan, *German Seed on Texas Soil*, 133 and n. 62.

[101] Cf. a similar reference to shingle-making in Ilse Wurster, Charles A. Kettner, eds., *The Kettner Letters*, 62.

[102] German settlers realized that European wines did not grow well on Texan soil. Typically, wine was made from wild grapes. The process of making wine from wild grapes differed from that of vineyard grapes in that it required exorbitant amounts of sugar after the initial fermentation. Cf. Terry G. Jordan, *German Seed on Texas Soil*, 140 and 77–79. In the mid 1870s, allegedly fifty thousand gallons of wine were produced in one season alone around Boerne in Kendall County. Cf. Terry G. Jordan, *German Seed on Texas Soil*, 140.

[103] One bucket (in German "Eimer"), was about 67.36 L.

[104] Name of a settlement in Blanco County, founded by John S. Hodges. Bergmann writes "Blanko."

[105] Probably people with whom he did gardening.

[106] Germans often sold fresh vegetables and melons at the San Antonio market because of the higher prices they could ask there. Cf. Terry G. Jordan, *German Seed on Texas Soil*, 136.

[107] A canister (in German "Kanne"), was about 0.93 L.

[108] Bergmann, as opposed to slaveholders, picks cotton himself.

[109] It seems that growing rye was difficult due to the climate and disease and often was not successful. Cf. Terry G. Jordan, *German Seed on Texas Soil*, 79–80. In a later letter Bergmann complains about an infestation with grasshoppers. See letter of 18 May 1867, where he refers to a failed wheat and rye crop due to grasshoppers. Nevertheless, his continued attempts to grow rye and wheat show his "grain preference brought from Europe." Cf. Terry G. Jordan, *German Seed on Texas Soil,* 130.

[110] A flail is a manual device used for threshing in order to separate grains from the husks. A manual threshing device usually consists of two or more large sticks, a long wooden handle or staff and a shorter, free-swinging stick attached to its end.
[111] The development of farm machinery did not affect the South as much due to the fact that the South did not grow the crops for which many of the machines were made. Germans, however, who grew a lot of wheat benefitted from this invention. Yet until the early 1870s, mechanical threshers were uncommon. Only in the mid 1870s was it common to implement farm machinery. Cf. Terry G. Jordan, *German Seed in Texas Soil,* 188. Later in a letter of 1881, Bergmann writes that his son Friedrich had bought a harvester and was going to buy a thresher. In a letter of 1885, he mentions that Friedrich had bought a threshing machine for 656 dollars the previous year and had already threshed close to 8,000 bushels.
[112] Tobacco was not unknown, but it was less common in Germany and raised only in small amounts in the southwest. Cf. Terry G. Jordan, *German Seed on Texas Soil*, 69 and 34. Of the Germans in Comal County less than 10 percent grew tobacco by 1860. Cf. Terry G. Jordan, *German Seed on Texas Soil*, 134. This is the only time Bergmann makes a reference to growing tobacco, so it seems that he later abandoned the idea.
[113] A small village a few miles north of Bergmann's home village of Ebersbach.
[114] Germans used many construction techniques that they knew from Germany, when they built their houses. Bergmann seems to have built a half-timbered construction (*Fachwerk*) that consisted of wood and sturdy stone. In the last decades of the 19th century, Germans discontinued stone constructions and instead built frame houses. Cf. Terry G. Jordan, *German Seed on Texas Soil,* 166–67. See also Randolph B. Campbell, *Gone to Texas*, 217.
[115] In German "Panna"
[116] In German colloquially "Ihr klötzelt."
[117] hundredweights
[118] From the German word "Neger," which at the time was not a slur against African Americans, but a term commonly used in Germany to describe a black person. Bergmann's comments on the conflict between the North and the South do not indicate that he had any racial biases.
[119] German settlers had formed political organizations in the area where Bergmann lived as early as 1853, such as *Demokratischer Verein, Der freie Verein*, and *Politischer Verein*. In a mass meeting in San Antonio in 1854, the German platform made what seemed to be a rather "ultra-radical" declaration that included the statement, "slavery is an evil, the removal of which is absolutely necessary according to the principles of democracy." S. Rudolph Leopold Biesele, *The History of the German Settlements*, 198–99 and n. 28. Obviously, not all German settlers felt equally and some felt one should not interfere with the affairs of the Americans (ibid. 202). Biesele also points out that any vote against secession during the Civil War was not necessarily opposition to slavery, but rather showed the wish to maintain the Union (ibid. 206).
[120] Again, this seems to be a monthly salary.
[121] Metaphorical image as a way to communicate sexual inactivity. The same image is used in a letter of 2 June 1857. See [3.1], p. 44.
[122] Gardening friends
[123] In the middle of the 19th century, monetary unity in Germany did not yet exist. Common currencies were the *Thaler* and *Gulden*. Bergmann probably refers to the *Vereinsthaler* that was introduced in 1857 and had replaced the standard Prussian *Thaler*. In Prussia and some other Northern states the *Vereinsthaler* consisted of 30 *Silbergroschen* with each *Silbergroschen* being 12 *Pfennig* (pennies). The *Vereinsthaler* also was the currency in Saxony from 1857 to 1873, where it also consisted of 30 *Silbergroschen*. However, each *Silbergroschen* was divided into 10 *Pfennig*. Later, in 1871, the Saxon *Vereinsthaler* was replaced by the *Mark* with 1 *Vereinsthaler* equaling 3 *Mark*.

[124] Probably the abbreviation for *Neugroschen*. In Saxony from where Bergmann came, the *Neugroschen* was equal to the Prussian *Silbergroschen*, but was divided into 10 *Pfennig*. The *Neugroschen* had been introduced to Saxony in 1841.
[125] In German "Unze." One ounce was about 29 g.
[126] In German "Prauschkurant," most likely a reference to the Prussian *Vereinsthaler*. Neugersdorf was a neighboring village of Bergmann's home village Ebersbach.
[127] It seems like Bergmann could not correspond with his relatives in Germany during the years of the Civil War. In contrast, Franz Kettner was able to manage to send letters by way of Mexico. For an interesting account of the situation in Texas during the Civil War, see his letter of 6 October 1863 and of December 1865, in Ilse Wurster, Charles A. Kettner eds., *The Kettner Letters*, 98–103; 112–119
[128] As a result of the Civil War, the ports on the Gulf of Mexico were blockaded. This also caused immigration to cease until the end of the Civil War in 1865.
[129] On 15 June 1864, Bergmann's own cattle brand was presented and recorded in the "Records of Marks and Brands Book E–1." See the records in the Kendall County Clerk's Office in Boerne, TX. Bergmann's cattle brand is "FB" on the left hip of the cattle, and ear marks with "half crop in right and split in left ear."
[130] In German "Nervenkrankheit." It is not clear what kind of illness Friedrich had; he might have suffered from typhoid fever, commonly called "nervous fever." During the Civil War, both Friedrich and Karl seemed to have been supporters of the Confederacy. Cf. Frank Wilson Kiel, *Civil War Soldiers in Kendall* County, 29; see also previous reference, p.86, n. 95.
[131] Boerne was a German settlement that was founded in 1852. Cf. Gilbert Giddings Benjamin, *The Germans in Texas. A study in immigration*, 117. It is the county seat of Kendall County and located thirty miles northwest of San Antonio. In 1856, a post office was created. For more information on Boerne cf. Vivian Elizabeth Smyrl, "Boerne, Texas," Handbook of Texas Online, <http://www.tshaonline.org/handbook/online/articles/hgb09>, [Accessed May 8, 2013].
[132] It is not clear to whom he refers, probably a relative.
[133] The name of Friedrich's wife was Johanna Bergmann. She was born on 14 April 1846 and died on 29 December 1931. They were buried next to each other in the Boerne Cemetery in Texas.
[134] In German "Bockanuß;" most likely, Bergmann's is referring to the bois d'arc tree.
[135] Bergmann explains an American measurement. One acre is 75 yards long and 75 yards wide, which equals 5,625 square yards.
[136] The valleys of the rivers in the western end of the "German Belt" typically had forests with cypress, pecan, and live oak trees, small oaks and junipers. Cf. Terry G. Jordan, *German Seed in Texas Soil*, 119–20. Franz Kettner mentions oak, elm walnut, and black walnut, cedar and cypress trees and also a type of resinous wood. Ilse Wurster, Charles A. Kettner eds., *The Kettner Letters*, 62.
[137] Orchards were part of German agriculture and that of the southerners. However, fruits like apples, pears and cherries generally did not grow well in the warm climate of Texas, so Germans largely abandoned growing them. Terry G. Jordan, *German Seed on Texas Soil*, 139. Apparently, Bergmann was more successful than others.
[138] Similarly, Franz Kettner writes he hopes that he will be spared being a soldier and that the war will soon end and that Texas will belong to the northerners again. He even explicitly states that the way Europeans are raised they are against slavery. See letter of 6 October 1863 in Ilse Wurster, Charles A. Kettner eds., *The Kettner Letters*, 98–99.
[139] I.e. the Confederacy
[140] In some counties, such as present-day Kendall County, unionism was strong. Cf. *German Seed in Texas Soil*, 182. Bergmann clearly was a Unionist. According to Helbich and

Kamphoefner, about 200,000 soldiers, or about ten percent of the Union army, were Germans. Cf. Wolfgang Helbich, Walter D. Kamphoefner eds., *Germans in the Civil War: The Letters They Wrote Home*, xi. In April of 1862, the Confederate Congress determined that all able-bodied white males between the ages of eighteen and thirty-five were eligible to be drafted; in September 1862, the upper age was raised to forty-five and in February 1864, to fifty, while seventeen became the lower limit. These laws intended to encourage volunteering. To be drafted was considered disgraceful. Cf. Randolph B. Campbell, *Gone to Texas,* 261.

[141] Bergmann confirms what Helbich and Kamphoefner pointed out. When asked by those who had remained in the homeland whether they should follow, usually many arguments for and against emigration were listed and it was pointed out that everyone had to make that decision for him- or herself. There was a mixture of joy about the possible arrival of relatives, but also worries about their chances of finding good jobs and a place to live. Finally, one did not want to be responsible for a decision that a relative might end up regretting. Cf. Helbich and Kamphoefner, *News from the Land of Freedom*, 29.

[142] This was a common complaint among German immigrants to Texas. Kettner also regrets that his next neighbor lives quite far away. See letter of 8 July 1860, in Ilse Wurster, Charles A. Kettner eds., *The Kettner Letters*, 94.

[143] No member of the Bergmann family seems to have visited relatives in Germany after their emigration.

[144] The word "zwieback" is German in origin and means "twice baked." It is a type of crisp, sweetened bread, made with eggs and baked twice. It is similar to Melba toast and has commonly been used to feed teething children or patients with an upset stomach. It probably was quite useful to people suffering from seasickness.

[145] A powerful disinfectant

[146] Ether alcohol (*Spiritus aethereus*) is a medicine traditionally used in patients affected by weakness, fainting, neuralgia, cramps, and vomiting. It was invented by the German doctor Friedrich Hoffmann in the beginning of the 18th century.

[147] Type of high-quality linen. The name "Atlas" is derived from the Arabic word for "smooth." Linen is made from fibers of the flax plant. When worn as garments, it is considered to provide exceptional coolness and comfort in hot weather.

[148] Type of fabric

[149] A centre-bit is a drilling bit with a central projecting point and two side cutters.

[150] A type of rifle with two barrels instead of one with which one can fire two quick shots; often used for hunting.

[151] Twill weave is one of the basic weave structures in which the filling threads are woven over and under two or more warp yarns, producing a characteristic diagonal pattern.

[152] The number probably refers to the diameter of the wool fiber that usually ranges between 16 μm and 40 μm.

[153] Pentecost is a Christian holiday on the fiftieth and last day of the Easter season. It celebrates the sending of the Holy Spirit to the Apostles. In 1871, it was on 28 May.

[154] Christoph married Maria Anna Reinhardt who was born in New Braunfels on 4 April 1952 and died on 17 October 1911. Christoph died on 18 May 1917. Their graves can be found in the Boerne Cemetery in Texas. Their marriage license was issued on 30 August 1871.

[155] Again, in German the word order is reversed, "I, Friedrich and Christoph." Kettner, in a letter of 15 March 1858, mentions that there is a saying in Texas that when someone has a child and it is a son that person has immediately earned $1,000, because a son will later help his father and, as a result, reduce the amount of money the latter has to pay by employing someone. Cf. Ilse Wurster, Charles A. Kettner eds., *The Kettner Letters*, 62.

[156] Horse theft by Indians seems to have been an ongoing problem in many areas. Franz Kettner in a letter of 26 July 1868, mentions the theft of two horses during his travels as a United States

assessor. Cf. Ilse Wurster, Charles A. Kettner eds., *The Kettner Letters*, 144. In a later letter, according to the editor, of November of 1869, Kettner writes that many Indians had been chased out of Kansas and into Texas and were making the region quite unsafe. In a harsh statement he expresses that the United States won't have a choice but to eradicate Indians altogether since in his eyes they were preventing the advancement of civilization. Cf. Ilse Wurster, Charles A. Kettner eds., *The Kettner Letters*, 162. In contrast to Kettner, Bergmann never uses derogatory terms for Indians.

[157] In 1871 and 1872, General Randall MacKenzie fought the Comanches. Finally, in 1874, the Red River War or Buffalo War began; this was the last Great Indian War in the Plains. The Red River War in Texas and the South Plains led to the "virtual extinction of the southern herd of buffalo, the final subjugation of the powerful Comanche." See "James L. Haley, "*Red River War*," Handbook of Texas Online, http://www.tshaonline.org/handbook/online/articles/qdr02, [Accessed May 8, 2013].

[158] See n. 164.

[159] Comfort was a township in Kendall County that was founded by German settlers in the 1850s. It is located about 50 miles northwest of San Antonio, Texas.

[160] This might have been an instance when Bergmann employed a German immigrant just like he himself had worked for someone else for a while upon his arrival in Texas, until he had earned enough money to do his own farming. See also n. 66.

[161] This is a good example for Bergmann's sense of humor and self-irony which seem to have helped him overcome difficult situations.

[162] In German, "I and Friedrich."

[163] In German, "I, Friedrich and Christoph."

[164] Texas had become the 28th state in the Union in 1845. In 1861, it had declared its secession from the United States to join the Confederate States of America. After the Civil War ended, the slaves were freed and Texas was part of the Reconstruction. In November 1869, Edmund J. Davis was elected Governor of Texas (See De Witt Clinton Baker, *A Texas Scrap-Book*, 297). He was a Unionist and a Republican, and remained Governor for four years; he was defeated in 1874.

[165] The German letter shows an interesting German spelling of the word 'Branch.' Bergmann writes 'Braensch.' This is another good example how the linguistic borders between German and English sometimes became blurred.

[166] In German "Elle," equals about 0,56 meter or 22 inches.

[167] Probably denim

[168] Oats were commonly grown in many areas of Germany. In Texas in the latter part of the 1800s, oats were often used as winter forage. Cf. Terry G. Jordan, *German Seed in Texas Soil*, 33 and 130.

[169] In German "nass," literally, 'wet.'

[170] Germans in the Hill Country produced large amounts of beef and pork and engaged heavily in sausage-making. This was a tradition that was basically unknown to the southern Anglo-Americans. Terry G. Jordan, *German Seed on Texas Soil*, 152–53.

[171] Friedrich buys land close to his father's. He pays $500 for the same amount of acres for which his father had paid $200 in 1857.

[172] Texas, like the other states in the South, lagged far behind in their transportation system. San Antonio did not have a transportation system until 1877, when it experienced a new time of economic growth due to the Galveston, Harrisburg and San Antonio Railway. Cf. T. R. Fehrenbach, "San Antonio, Texas," Handbook of Texas Online, <http://www.tshaonline.org/handbook/online/articles/hds02>, [Accessed May 8, 2013]. Franz Kettner, in a letter of March 15, 1858, mentions that a railway is being built that connects Indianola and San Antonio. Cf. Ilse Wurster, Charles A. Kettner eds., *The Kettner Letters*, 62. Railroad service from Indianola to

the interior of Texas began in 1871. Cf. Brownson Malsch, "Indianola, TX," <http://www.tshaonline.org/handbook/online/articles/hvi11>, [Accessed May 12, 2013].

[173] Germans in Texas had a tendency to form clubs (*Vereinswesen*), of which only Germans were members. Typically, during meetings German was used. There were farming, shooting, singing, literature, gymnastics, and mutal aid clubs, but most of them were gone by the end of the 19th century. Cf. Terry G. Jordan, *German Seed on Texas Soil*, 106. Almost every German settlement had singing societies the first one of which, the "Gesangverein Germania," was formed in 1850 in New Braunfels. Cf. R.L. Biesele, *The History of the German Settlements in Texas*, 222. The first German-Texas *Sängerfest* (song festival) was held in New Braunfels in October of 1853. Cf. R. L. Biesele, *The History of the German Settlements in Texas*, 217. Singing societies from San Antonio, Sisterdale and New Braunfels participated in this festival. A *Sängerfest* was at times also a place where political ideas were expressed. Cf. Gilbert Giddings Benjamin, *The Germans in Texas*, 117. The second German-Texas *Sängerfest* (singers' festival) was on May 13, 1854 in San Antonio. In 1854, the state-wide singing society "*Deutsch Texanischer Sängerbund*" was established. Cf. Andreas V. Reichstein, *German Pioneers on the American Frontier. The Wagners in Texas and Illinois*, 77–78.

[174] Obviously, Bergmann must have heard news about his sister that upset him.

[175] One fluid quart equals 0.946 L.

[176] Bergmann probably made his wine from wild grapes and seems to have been able to produce wine in spite of the drought in 1879.

[177] Word or perhaps line missing

[178] Karl's second wife was Pauline Vogt, born on 30 November 1856. Their marriage license was issued on 4 November 1881. Pauline died on 27 January 1934. Karl died on 21 November 1938 at the age of ninety-three. Their graves can be found in the Boerne Cemetery in Texas.

[179] In 1858, both Fredericksburg and Comfort formed singing societies. Cf. R. L. Biesele, *German Settlements in Texas*, 223.

[180] In October 1881, the small settlement of Boerne established the "Gebirgssängerbund von West Texas" [literally 'The High Elevation Mountain Singing Society of West Texas']. During the song festival at the time, among the participating societies were: "Liedertafel" from Comfort, "Gesangverein" and "Gemischter Chor" from Boerne, "Concordia" from Fredericksburg, and "Walhalla" from Comal County. The program had some nineteen songs, among them the *Bundeslied* by Mozart, the *Jäger's Abschied* by Mendelsohn, and the *Vaterlandslied* by Marschner. Cf. Gilbert Giddings Benjamin, *The Germans in Texas. A study in immigration*, 118.

[181] Word omitted in the original letter, possibly, the word "rich." However, it cannot be ruled out that it was a word that further described a poor person.

[182] In German literally "everyone is wearing the same shoes."

[183] The currency of the German Mark was the result of a currency reform following the unification of Germany under Bismarck in 1871.

[184] A German newspaper

[185] Bergmann's spelling in German of 'spring' is "Früjar" instead of 'Frühjahr,' probably as a result of lack of practice in writing in German.

[186] Dolomite and limestone were common kinds of stone in the area. Cf. Vivian Elizabeth Smyrl, "Kendall County," Handbook of Texas Online, <http://www.tshaonline.org/handbook/online/articles/hck03>, [Accessed May 8, 2013].

[187] In German "Grüβend" is the gerund of the German verb 'grüβen,' in English 'to greet.' This is an Anglicism that cannot be used in proper German.

[188] This is a comment clearly intended as a funny note. It is on the bottom of the page of the original letter, in which Bergmann also uses the German word for 'not' in dialect, "nischt" as opposed to 'nicht.'

[189] Bergmann was a member of the Germania Farmer Association. Many Germans on the entire state of Texas formed societies that were devoted to agriculture in an attempt to improve farming methods. The *Germania Farmer Verein* (German Farmer Association) was an association founded in 1875 with the purpose to protect cattle from thieves and Indians. The cattle were branded by the residents with the association brand G and their efforts were successful. The association received a state charter by 7 February 1876. Cf. David DeKunder, "Anhalt Hall," Handbook of Texas Online, <http://www.tshaonline.org/handbook/online/articles/xda04>, [Accessed May 8, 2013]. The original of this document is written in pencil.
[190] I.e. 24 November 1888
[191] This was a German-language paper founded in San Antonio by the German immigrant August Siemering in July of 1865. The *Freie Presse* was published until 1945 when publication ceased. Cf. "German-language paper founded in San Antonio," Handbook of Texas Online, <http://www.tshaonline.org/day-by-day/31313>, [Accessed May 8, 2013].
[192] The newspaper was founded in 1851 by German immigrant Friedrich Hermann Seele and Ferdinand Jacob Lindheimer, the great Texas botanist. Cf. Edward C. Breitenkamp, "Seele, Friedrich Hermann," Handbook of Texas Online, <http://www.tshaonline.org/handbook/online/articles/fse06>, [Accessed May 8, 2013].

Plates

Plate 1: Christian Friedrich Bergmann *Plate 2*: Johanna Christiane Luise Bergmann
Courtesy of Patrick Heath Public Library Archives in Boerne, Texas

Plate 3: Bergmann Family Estate in Ebersbach, Germany (sold in 1876)
Courtesy of Helga Mittrach

Plate 4: Family of Karl Heinrich and Pauline Bergmann (front). From left: Louis Adolph, Otto Henry, George William, Emelia Luise (maiden name Potchernick), Ella (maiden name Agold). Courtesy of Patrick Heath Public Library Archives in Boerne, Texas

Plate 5: Family of Friedrich and Johanna Bergmann (ca. 1892). From left: Ernest, Max, and Fritz Bergmann. Far right: Augusta Bergmann with husband and son Ernst. Front row from left: Friedrich Bergmann and Johanna Bergmann (maiden name Menzel) Courtesy of Patrick Heath Public Library Archives in Boerne, Texas

Plate 6: The Bergmann Family Letter Collection
Left: First letter from San Antonio (1856)

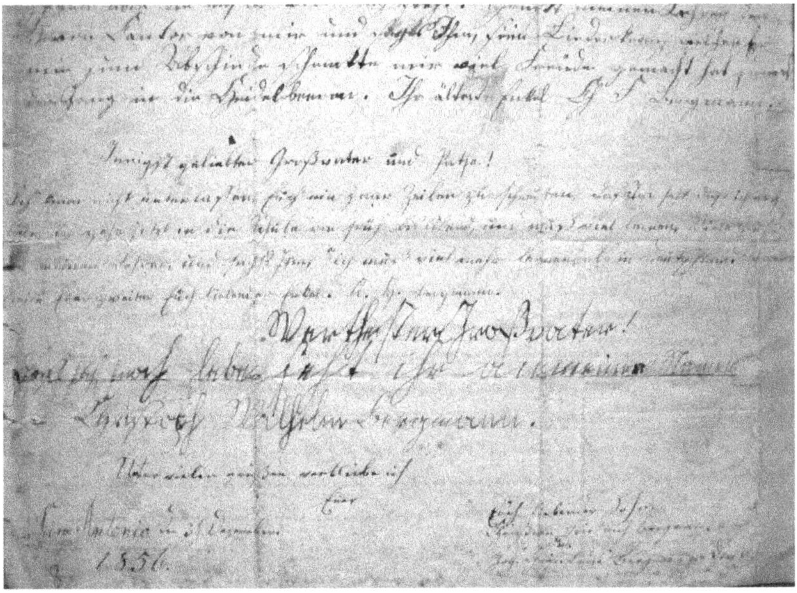

Plate 7: Notes and signatures of the Bergmann sons, from oldest to youngest, in the first letter from San Antonio (1856)

Plate 8: Grave stones of Christian Friedrich and Johanna Christiane Luise Bergmann near Boerne, Texas. Photograph by Gene Steele (2013)

Plate 9: Bergmann's land with fenced-in grave site of Christian Friedrich and Johanna Christiane Luise Bergmann in background. Photograph by Ruth Cape (2013)

Plate 10: Graves of Christian Friedrich Bergmann (son) and his wife Johanna.

Plate 11: Graves of Christoph Wilhelm Bergmann and his wife Anna

Location of both graves: Boerne Cemetery, Texas. Photographs by Gene Steele (2013)

Plate 12: Graves of Karl Heinrich Bergmann and his second wife. Boerne Cemetery, Texas Photograph by Gene Steele (2013)

Plate 13: Old stone fence on Bergmann's land near Boerne, Texas. Photograph by Ruth Cape (2013)

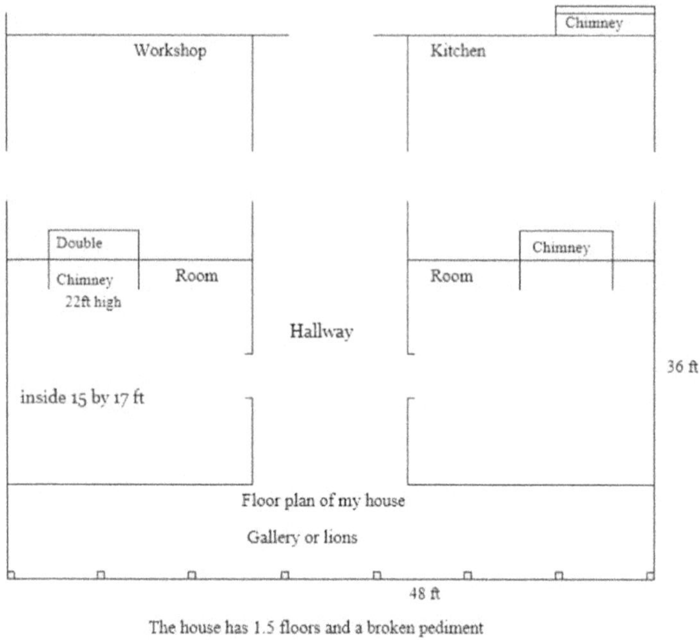

Plate 14: Floor plan of Bergmann's house included in a letter of 28 June 1866. The drawing was made by one of his friends. (cf. letter 8.1). Reproduction of the original drawing by Lauren Tepera. *Plate 15* below: Original drawing

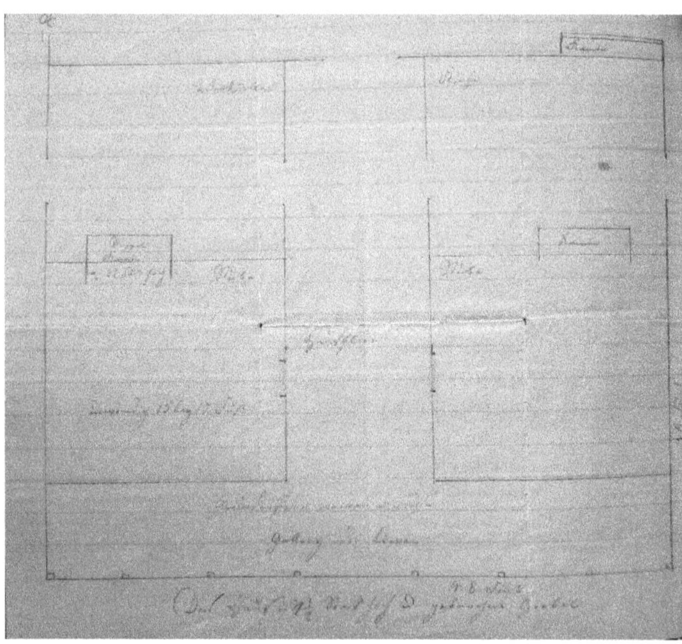

Part Three:

DIE BERGMANN-BRIEFE

1. Die Seereise nach Amerika: August bis November 1854

[1.]
[Beschreibung der Seereise nach Amerika 1854; ohne Datum, wahrscheinlich von Ende 1854/Anfang 1855]

Lieber Vater!
Mit Zittern und Zagen ergreife ich die Feder! Ob Euch mein Brief noch am Leben treffen wird oder nicht? Ihr werdet mit Sehnsucht einen Brief von mir erwartet haben, weil ich Euch versprach, gleich zu schreiben, sobald ich auf festem Boden stehe; aber leider, ein Brief kostet viel und ich wußte, Heuchling hatte geschrieben, so hoffte ich, Ihr würdet es erfahren, daß ich und meine Familie glücklich und gesund an Land gekommen sind.

Jetzt will ich Euch Nachricht über meine schwere und gefahrvolle Reise mittheilen. Den ersten Tag nach meinem Abschiede kamen wir nach Leipzig und blieben die Nacht im Schwarzen Kreutz. Lieber Schwager, da wirst Du wissen, wie es in Leipzig ist, zumal mit Auswanderern, weil die Geld haben— und überall bis nach Bremerhaven. Den zweiten Tag blieben wir in Braunschweig, wo grade Messe war. Den dritten Tag kamen wir nachmittags vier Uhr nach Bremen. Dort mußten wir warten bis zum sechzehnten [August]. Wir fuhren dann mit dem Dampfschiff *Roland* nach Bremerhaven, wo wir ins Einwandererhaus gehen mußten. Unterwegs fuhren wir uns bei Vegesack fest, weil gegen 500 Passagiere darauf waren. Wir standen 3 Stunden fest. Erst gegen 7 Uhr fuhr es wieder fort, weil die Fluth eintrat, und so kamen wir abends 11 Uhr nach Bremerhaven, wo uns ein sehr hartes Lager begrüßte.

Am 19. wurde unser Schiff *Minna Blumenthal* durch das Dampfschiff *Magnat* aus dem Hafen bugsiert und wir lagen, wegen Jahrmarkt, bis zum 21. vor Anker. Den 21. gegen 10 Uhr wurden die Anker gelichtet und wir segelten bei gutem Winde mit vollen Segeln in See. Das Interessante, wie sich die Weser und Ostsce [Nordsee] miteinander streiten, so daß diese beiden Elemente in ein Scheinen geriethen, da haben wir 3 Sorten Wasser geschen. Am 22. wurde der Wind stärker, so daß er sich am 25. fast zum Sturm erhob. Gegen Abend wurde er so heftig, daß die Kisten im Zwischendeck übereinander stürzten und Christoph unter Seilers Kommode kam, daß wir glaubten, es wären beide Beine zerschlagen. Auch Karl hatte einen Stoß davongetragen, denn die Beiden mußten aufwarten, weil wir alle Drei durch das Schaukeln die Seekrankeit hatten; gegen Mitternacht ward es ruhiger, und am 26.–27. trat Windstille ein.

Am 28. ging ich früh vor Sonnenaufgang aufs Deck—ein schöner Morgen. Die Sonne strahlte dann so freundlich, und wir waren im Kanal. Die Kreidefelsen ragten von der englischen Küste gleich den Sächsischen Schweiz-Klippen empor. Von früh 8 Uhr bis 3 Uhr Nachm. war Windstille. Gegen Abend kam ein englisches Kriegsschiff. Auch zählte ich noch im Kanal 61 Fahrzeuge. Um Mitternacht sah ich links vom Verdeck aus eine englische Stadt, die wie in der schönsten Illumination vor meinen Augen lag. Am 29. früh kam ein kleiner Vogel, wie ein Zeißig, am Tauwerk hinauf, welcher die Insekten auffraß. Auch hörten wir morgens wie abends einen von der französischen Küste ähnlichen Kanonendonner.

Den 30. früh sahen wir einen englischen Schraubendampfer, welcher immer im Kanal herumsegelte und, wie der Kapitän meinte, ein Beobachtungsschiff sey. Auch sahen wir eine Menge sehr große Fische unter dem Namen „Stier". Gegen Abend sahen wir eine Seespinne, welche bei ihrem Auftauchen einem Todtenkopfe glich; so auch einen Zug kleine Fische. Im Kanal hatten wir wenig Wind. Erst am 31. Mittags bekamen wir Morgenwind, wo gleich die ganzen Segel gespannt wurden und uns unsere *Minna* dem Ozean schneller entgegentrug. Am 1. Sept. hatten wir noch guten Wind und gegen 8 Uhr früh sahen wir wieder die englische Küste, welche aus Gebirgen bestand, die zum Theil holzbewachsen, zum Theil urbar waren. Auch sahen wir Felsen, auf denselben Leuchtthürme sowie auch Städte und Dörfer liegen, und bei Sonnenuntergang kam ein schwedisches, nach Ostindien fahrendes Viermasterschiff, und wir verließen den Kanal. Den 2. hatten wir guten Wind, zwei Sorten Fische sahen wir, Schweinefische und Nordkapp. Letztere, meinten die Matrosen, wären zu groß für die Harpune zum Fangen. Den 3. früh kamen zwei Hayfische, welche kleine Fische (Delphine) zur Begleitung hatten, gegen 01 Uhr kam von Westen eine französische Brigg, welche von Afrika kam, aus der ein Kahn mit vier Matrosen auf unser Schiff kam, um Lebensmittel zu holen. Denn sie waren vom Sturm verschleudert worden. Heute abend starb eine Frau, welche schon krank an Bord kam. Sie wurde in einen Strohsack genäht, welches die Matrosen unternahmen, und künftigen Abend unter dem Gebet der wogenden See übergeben. Am 4. überfuhren wir eine Brigg, auch ein Sturmvogel kam aufs Tauwerk in Gestalt einer Lachtaube. Den 5. zwei große Schweinefische, vom 1.–5. [September] war Südostwind, den 6. drehte sich der Wind nach Westen und kam uns auch noch günstig.

Vom 3.–8. [September] war meine Frau sehr krank. Sie hatte furchtbare Kopf- und Unterleibsschmerzen, so daß es der Ruhr glich. Den 7. war der

Wind nicht günstig, den 8. wurde ich sehr krank, wo ich Fieberanfälle und furchtbare Kopf- und Unterleibschmerzen hatte und erst am 13. wieder anfing, etwas zu essen. Während der Krankheit habe ich nicht geschrieben, ist auch nichts vorkommen als Zänkerei und sogar Schlägerei. Den 17. starb ein neugeborenes Kind. Diesen Abend passierten wir die Kanarieninseln, wir haben sie aber nicht gesehen. Den 18. starb wieder ein Kind an Zahnfieber, 5 Monate alt, und auf den Abend wurde es dem Ozean übergeben.

Von da hatten wir guten Wind bis zum 23., wir fuhren in vier Stunden 10 deutsche Meilen. In dieser Zeit haben wir verschiedene Fische gesehen, Fliegende, Delphine, usw. Den 23.–26. Windstille, denselben Tag sahen wir eine Brigg links. Am 27. zwey Schiffe, welche uns entgegenkamen, eine Brigg und eine Barke. Wir sahen große Fische, Speerhai genannt, immer 5 bis 6 Ellen lang. Den 29. gegen Abend kam ein großer schwarzer Vogel ähnlich einem Fischreiher, den 30. kam ein großer schwarzer Vogel, wie ein Rabe. Den 1. Oktober noch guten Wind. Abends wurde zum zweiten Male getanzt.

Vom 18. [September] an war es, nähmlich bei schwachem Wind, sehr heiß. Ich muß bemerken, daß wir bis jetzt wenig Regen gehabt haben. Regnete es an einem Tage, so hatten wir auch Sonnenschein. Die Witterung ist auf der See ganz anders als auf dem Lande. Wir wünschten uns manchmal Regen wegen dem Trinkwasser, da gab's sehr wenig, für 5 Personen nicht gut 3 Seidel und was für Wasser! Da gedachte ich an den Brunnen beim Teiche und der hinteren Wiese, aber leider es war nicht. Die Sonne steht in der Nähe von Kuwa (Kuba) fast senkrecht, daher ist es so heiß, das Wasser wird in den Fässern warm und stinkig. Unseren 3 Jungen schmeckte es zwar besser als zuhause das beste Quellwasser. Christoph hatte es die letzte Zeit besser; er hatte sich bei den Matrosen beliebt gemacht, auch beim Koch. Sogar der Kapitän wollte ihn haben, sogar kaufen, weil er selbst kein Kind hatte.

Übrigens muß ich bemerken, nach einer solchen langen Fahrt sehnten wir uns alle nach festem Lande, wie ein Fisch nach Wasser, der in der Wärme stundenlang getragen ist. Den 5. [Oktober] wurde ein Dieb ertappt, welcher Kleinigkeiten entwendet hatte, der Kapitän faßte ihn beim Arme, und der Bootsmann (Untersteuermann) legte ihm mit gesponnenem Hanf 10 auf den Hintern. Abends waren auf beiden Seiten Gewitter, erst am 6. früh regnete es, so wurden die Wasserfässer mit dem Wasser gefüllt; auch wir füllten alles was wir hatten; das half einige Tage für den Durst. An diesem Tage sahen wir einen sehr schönen Schmetterling. Den 7. Windstille und sehr große Hitze, daß im Schatten 28 und in der Sonne 38 Grad Wärme war. Am 10. bis Abends 12 Uhr Gewitter; denselben Tag sahen wir sehr schöne Wasservögel von verschiedener Gattung. Den 11. kam eine Schwalme, ähnlich Euren

Feuerschwalmen und begleitete uns mehere Tage. Den 12ten Abends Donnerwetter mit starkem Wind, am 13. wieder Donner.

Den 14ten [Oktober] sahen wir zwei Inseln Tigua und Monserrat, welch letztere unseren Augen einen herrlichen Anblick bot, denn wir fuhren ziehmlich nahe vorbei. Unten am Strande war ein Städtchen, auf einigen Bergen war Tabak gepflanzt. Auch sahen wir durch das große Feuerrohr Palmenbäume und hie und da Häuser, worüber sich alle Passagiere königlich freuten. Abends waren ringsum Gewitter, denn wir waren nur 16–17 Grad vom Äquator entfernt. Unser Kapitän machte einen Umweg von 100 deutschen Meilen.

Den 15. früh wurde ein totes Kind geboren, den ganzen Tag Regen. Den 16. starb ein zweijähriges Kind, welches die ganze Reise kränklich war. Auch wurde ein fliegender Fisch gefangen. Der Körper sah fast vierkantig aus. Den 16. erblickten wir, wegen ungünstigem Winde, Puertoriko. Am 17ten kamen mehrere Flüge Zugvögel, die so hungrig waren, daß sie, während ich mit den Kindern das Mittagessen verzehrte, einer auf meinem Kopfe, der andere auf der Schüssel saß, und blieben mehrere Tage an Bord.

Den 18ten starb ein sechsjähriger Knabe an einem nervischen Fieber, und auf den Abend den schäumenden Wellen übergeben wurde [sic]. Den 20. wurde ein Schwimmvogel geschossen von der Größe einer Ente, der sich vorn aufs Bugspriet setzte. Vormittags sahen wir Domingo, und erst am 24. verschwand sie unseren Augen, auf welcher ein Berg von 6000 Fuß über die Meeresfläche hervorragte und vom Schiffe sehr gut zu sehen war. Den 21. starb ein Mädchen von zwei Jahren an Zahnfieber. Auch sahen wir wieder sehr große Fische, Ungeheuer genannt. Den 23ten fuhren wir vormittags 9 Uhr an einem verunglückten Schiffe vorbei, welches jedenfalls wenige Tage vorher durch den Orkan bei Calverston [Galveston] umgeworfen war, wie der Kapitän meinte, solches in diesen Zonen vorkäme. Er sagte, es wäre ein Dreimaster, deshalb er auch jedesmal bei Gewitter die Segel einraffen liess. Ihr könnt Euch wohl denken, welch Schauern uns alle ergriff, als wir dieses erblickten.

Den 24. [Oktober] Mittags sahen wir Jamaika, welches eine Länge von 40 geographischen Meilen enthält und sich mit einer großen Gebirgskette unseren Augen präsentierte und bei welchem wir am 25ten Nachmittags vorüber segelten. Den 27. starb wieder ein Kind, 1½ Jahr alt, an Zähnen und hernach Geschwulst. Auch kamen wir den Morgen zur Südspitze von Kuwa [Kuba], Kobantan genannt, wo ein kolossaler Leuchtthurm stand, bei welcher wir mittags schon vorüber fuhren und dann in den Golf von Mexiko kamen. Am 28ten starb Heuchlings jüngstes Mädchen. Den 29.–30 hatten wir Windstille

Den 31. Oktober und 1.–2. November hatten wir sehr guten Wind, so daß wir am 3. früh Calverston sahen; der Tag zuvor war mir, als verkündigte es ein groβes Fest. O welche Freude.

Bei Sonnenaufgang sahen wir Calverston [Galveston]. Alles jubelte und jauchzte vor Freude, nach solcher langer Wasserfahrt festen Boden zu betreten. Gegen 8 Uhr wurden die Anker geworfen, weil noch kein Lotsen am Bord war. Denn es sind da sehr grosse Sandbänke, welche die Fahrt gefährlich machen. Gegen 9 Uhr kam der Lots an Bord, welcher uns nahe am Hafen brachte, aber leider, wir durften nicht landen, weil das gelbe Fieber noch in der Stadt war, und so lagen wir bis zum 5. vor Anker. Denselben Tag fuhren wir Mittags mit dem Dampfschiff fort, welches uns am 6. mittags nach Indianola brachte.

Übrigens muß ich bemerken, daß wir es auf unserer *Minna* mit dem Essen sehr schlecht hatten, denn mit 7 Wochen bekamen wir nur das halbe Brot, drei Wochen lang alle Abende Reissuppe ohne Brühe. Unser Kapitän war sehr geizig, sonst ein guter Mann, auch die anderen Matrosen sehr hübsche Leute. Er hatte das Schiff auf 110 Mann verproviantiert und 150 waren wir drauf, weil er aus dem Hafen hinausgerückt war, nahm er immer noch ganze Kähne voll darauf. Hingegen hatten es die Leute auf dem Schiffe *Neptun* und *Franziska*, welche 14 Tage später in Bremen abgefahren waren und mit uns zugleich die Anker warfen, sehr gutes Essen und Brod genug.

Zwei Tage blieben wir dann in Indianola, dann fuhren wir fort mit fünf Mauleseln. Da waren wir vier Parten zu der Ladung, unser Fuhrmann hieß Schlein. Den 1. Tag passierten wir lauter Prärie. Den 3. Tag schossen wir zum Abendbrot eine wilde Gans, die sogleich gerupft und gebraten wurde, und den 4. Tag abends kam Norder (sehr kalter Nordwind) mit Eis. Da sind wir bald erfroren, wenn ich Betten nicht hatte, denn wir mußten die ganze Reise unter freiem Himmel schlafen. Den 5. Tag ruhten wir wegen der Kälte nicht weit von einem Farmer; das war am Sonntag. Nun haben wir unsere Reise wieder fortgesetzt. Den 2ten Tag nach dem Norder kamen Heuschrecken, daß alles voll war, über und über, so daß sie uns die Augen bald ausflogen auf der Ruth (Road = Straβe). Wir kamen nun den 18. [November] in San Antonio an. Da sind wir 8 Tage bei Opitz aus Löbau geblieben. Dann haben wir uns selbst ein Quartier gerent (=gemietet), einen Monat 7 Dollar.
[Hier bricht der Originalbrief ab.]

2. Die ersten Jahre in San Antonio: 1854 bis 1857

[2.]
[San Antonio, den 3. Dezember 1856: Das Titelblatt des Originalbriefs zeigt in Rahmenanordnung Zeichnungen mit Ansichten von San Antonio: Oben ein Bild mit dem Titel „Main Plaza, San Antonio, Texas“. Darunter eine Girlande von Blättern und Obst und die englische Beschriftung: „Drawn after Nature by Erhard Pentenrieder. Published by Pentenrieder & Blersch, San Antonio, Tex“. Links auf der Titelseite vertikal angeordnet Bilder mit Betitelung von der Mission San José, Mission Conception, dem Gerichtsgebäude und einem Reiter auf einem heubepackten Esel; rechts vertikal angeordnet Bilder des Alamo, der Mission San Juan, einem Wohnhaus mit zwei Personen davor und einem Cowboy, der mit einem Lasso einen Büffel einzufangen versucht. Ausserdem Abbildungen oben links von einem stehenden Indianer und oben rechts von einem stehenden Mexikaner. In der Mitte des Randes links eine Abbildung von einem Schwarzen, in der Mitte des Randes rechts eine Abbildung von einem Amerikaner, beide in sitzender Stellung, der Amerikaner in nachdenklicher Haltung, das Kinn in eine Hand gestützt und mit der anderen ein Gewehr haltend. Außerdem zwei Wildtiere, links ein klettender Bär und rechts eine Wildkatze mit bedrohlicher Gebärde.]

Lieber Vater!
Ich kann nicht unterlassen, Euch wieder zu schreiben. Es kümmert mich doch, was Ihr und meine Geschwister machen. Seid Ihr noch alle gesund und munter? Ich habe Euch schon drei Briefe geschrieben, und Ihr habt mir noch keine beantwortet. Im ersten habe ich Euch unsere ganze Reise beschrieben und auch schon das, was Ihr gern wissen wolltet von Heuchling; ich und meine Frau wollten nicht Gesinde sein. Gleich nach dem ersten Briefe erhielt ich einen Brief von Euch, welcher mir ganz unerwartet kam und sehr viel Freude machte, wie ich gelesen habe, daß Ihr noch alle gesund waret. Darauf habe ich gleich wieder einen geschrieben, ich habe aber einen Monat um den anderen vergeblich gehofft auf Antwort. Eines Sonntags kam Bundesmann zu mir und sagte: Dein Vater ist am letzten Dezember bei meinem Vater gewesen. Da habe ich [er] gefragt, ob er nichts von dir weiβ, was mich sehr angriff, woraus ich schliessen konnte, daß Ihr keinen meiner Briefe erhalten hättet. Den dritten Februar schrieb ich wieder einen Brief, in welchem ich auch schon um das Geld gebeten habe, weil Ihr mir geschrieben hattet, daß Ihr mein Haus verkauft

hättet. Auch hatte ich in denselben einen Golddollar zwischen den Siegel gelegt, welcher als Pathengeld gelten sollte, für Karle seinen Karl, weil er mir noch den letzten Tag versprochen hatte, einen Gevatterbrief zu schicken. Auch hat Bundesmann zu gleicher Zeit an seinen Vater geschrieben und auch meinen Brief mit besorgt, aber er hat auch noch keine Antwort. Ich ersuche Euch nochmals dringend, schickt mir das Geld, denn ich will mir Land kaufen. Meine Mittel erstrecken sich noch nicht so weit, daß ich mir eine Farm kann einrichten, denn es gehört schon viel dazu, eine Farm einzurichten. Die Lebensmittel und Rente [Miete] sind theuer, und ehe man selbst ernten kann, da soll schon reichlich Geld sein. Sobald ich aber das Geld habe, will ich mir gleich eine Menge Kühe kaufen, so daß ich von denselben kann leben, und ich hoffe, daß Ihr mir dasselbe sobald als möglich schicken werdet.

Wir sind alle Gottlob noch gesund, Karl und Christoph gehen jetzt noch ins Kloster in die Schule, wo sie zur besten Zufriedenheit lernen. Friedrich dient jetzt vier Monate, er hat sechs bis sieben Dollar und Essen. Sonst gingen alle drei in die Schule, welches sehr viel Geld kostet. Ich muß für jeden den Monat 1 und ½ Dollar Schulgeld geben, lernen Deutsch, Englisch und auch etwas Spanisch, sind aber den ganzen Tag in der Schule. Es ist ein lutherischer Priester hierhergekommen, welcher alle Sonntage predigt, welches bisher nicht der Fall war, und so will ich auch Friedrichen, sobald es geht, konfirmieren lassen. Uns hat es anfangs bange gethan, denn von Evangelischem Gottesdienst war keine Rede, mehr von dem: heute ist einer erschossen oder erstochen worden; denn schon auf den Bällen sind Schuß-Hotter häufig Mode, ...[unleserliche Zeile], so ist der Schuß-Hotter der dritte oder vierte Mann, denn hier gilt ein Menschenleben nicht mehr als in Deutschland das Leben eines Hundes, denn so recht strenge Untersuchung ist nicht, zumal wenn es das Leben eines Deutschen oder Irländers betrifft.

Ich bin eben dieses Jahr wieder einige Monate in der Prärie gewesen und habe gemäht. Da habe ich täglich einen Dollar und die Kost, später habe ich für mich gemäht und einen Heustock gemacht. Den will ich diesen Winter verkaufen. Meine Frau wäscht und bügelt für die Leute, auch für Bundesmann und andere Ladendiener, wo sie für das Dutzend einen Dollar hat, und damit hat sie sich diesen Sommer schönes Geld verdient. Dieses Jahr war ein sehr schlechtes. Der Winter war so kalt, daß es keinen Mexikaner gedachte, so dickes Eis gesehen zu haben. Und ebenso heiß war auch der Sommer, daß Korn und die Bathaten (süße Kartoffeln) sind sehr schlecht gerathen. Deutsche Kartoffeln wachsen um San Antonio keine, denn der Boden ist zu fett. Am besten wachsen sie auf sandigem Boden. Drei Pfund Kartoffeln kosten 25

Cent, Weizenmehl 12 Pfund einen Dollar, Maismehl 1 Dollar 30 Cent, und so ist eben alles dieses Jahr sehr theuer.

Jetzt, lieber Vater, habe ich mir ernsthaft vorgenommen, Land zu kaufen. Denn wenn ich auch etwas mehr für den Acker geben muß, ich will nur darauf sehen, daß ich Holz und Wasser habe und nicht den Indianern an ihre Grenzen komme; denn wenn sie einen Weißen können tödten, da meinen sie, sie thun Gott einen Dienst. Überhaupt stehlen sie gern Pferde, Maulthiere, Ochsen und Kühe. Ich habe jetzt 5 Stück Vieh, 2 Kühe und 3 Mutterkälber, wovon mir zwei nächstes Jahr auch wieder Kälber machen, auch 20 Hühner. In San Antonio kann man rechnen, daß jährlich mehr als 150 neue Häuser gebaut werden, und doch ist für mich diesen Winter wenig Arbeit, denn es sind hier nur Tischler, Maurer und Kaufleute, und wenn Arbeit ist, so sind die Polen; diese arbeiten billiger als die Deutschen und [daher] habe ich fest vorgenommen zu farmen.

Lieber Schwager, Du hast mich in dem Briefe gebeten um Blumensamen, aber leider, das läßt sich hier um San Antonio sehr schwer möglich machen, denn es ist zu viel Vieh hier, welches schon die Blumen abfrißt. Wenn ich einmal weiter von der Stadt bin, dann läßt sichs besser machen. Obst, nähmlich Birnen, Äpfel, Pflaumen, Kirschen sind wenig, wilde gibt es von allen Sorten. Pfirsichbäume sind bis jetzt die einträglichsten, denn wenn man Steine, nähmlich Pfirsichsteine legt, so tragen sie schon oft im zweiten Jahr. Ich habe für mich 2000 Pfirsichsteine gelegt, auch will ich noch etwas Apfelkörner legen, mein Wirth hat mir Land im Garten gegeben für den Samen. Ich habe auch voriges Jahr sehr viel Schweine geschlachtet und auch dieses Jahr wieder einen Anfang gemacht.

Ich bitte Euch, lieber Vater und Geschwister, schreibt mir doch alle, was Ihr macht. August, hast Du Dich noch nicht vermehrt? Und Luise, hast Du noch nicht geheiratet? Wenn das nicht der Fall ist, so schreib mir, daß ich Deine Hochzeit auch in Texas feiern kann, freilich, das Kalb kann ich Dir nicht führen.

In San Antonio wird Theater gespielt von den Deutschen, auch Bundesmann und ein Löbauer, Becker, spielen mit, worunter Bundesmann gewöhnlich die beste Rolle spielt. August, da wärest Du eben auch brauchbar, um die Rolle des dummen Jungen zu spielen. Aber thut mir auch den Gefallen, grüßet mir meine alten Nachbarn, vorzüglich Wendebaum und den Stellmacher, aber auch meine Blumenkollegen, namentlich Neumann sowie auch Tischlerkarl, Obstel, Langkreißel, und was ich muß auch bemerken, daß Ihr meinem Gevatter Bergmann auf der Haine Notiz von meinem Befinden gebet.

Lieber Vater, wenn Ihr mir das schickt, so müßt Ihr womöglich New Yorker Wechsel kaufen, weil hier die Kaufleute die mehrsten Geschäfte dorthin machen. Auch habe ich noch heute mit Bundesmann gesprochen, daß, wenn es zu viel Schwierigkeiten machen sollte mit den Wechseln, er mir das Geld auszahlen wollte, sobald er den Brief von seinem Vater erhalten habe. Auch er wollte heute Abend noch schreiben an seinen Vater, daß er, sobald er wüßte oder wenn er das Geld hätte, so schleunigst als möglich schreiben sollte. Lieber Vater, so regelt mir die Sache, denn es wird jedenfalls eine gerichtliche Schrift nöthig sein, damit ich Bundesmann kann Gewißheit darüber geben, ob Ihr das Geld an den alten Bundesmann ausgezahlt habt oder ob er es auf sicherer Hypothek bestehen lassen will.

Grüßt alle meine Freunde von mir, die ich nicht alle nennen kann, und dann sagts ihnen, ich befinde mich noch wohl. Freilich, Texas wird kein Deutschland; obgleich ich nicht soviel Arbeit habe, so hat man doch auch immer viel Sorgen. Denn wenn der Monat um ist, müssen wir 5 Doll. Miethe zahlen, und wir haben nur ein kleines Zimmer, welches immer noch billig ist. Die Arbeit wird zwar gut bezahlt, von 75 Cent bis 1–2 Doll., aber was bei Euch fünf Pf[ennig] kostet, ist hier öfter 5 Cent.

Auch muß ich bemerken, Diebe und schlechte Menschen, auch Trunkenbolde, giebt es genug (Laufers genannt), die von keiner Arbeit nichts wissen wollen. Die Stadt ist jetzt ziehmlich groß und allerlei Volk darin aus allen Ländern, die Ihr nur denken könnt; sogar die Araber fehlen nicht mit ihren Kamelen. Viele Deutsche sind wegen Betrug aus Deutschland fortgemacht. Die bringens hier auch noch fertig, weil die Amerikaner desgleichen auch thun. Zutrauen darf man keinem schenken. Das Geld bringt ziehmlich gute Zinsen. Wer welches hat, das Hundert 2 Dollar pro Monat, das ist das Wenigste. Das ruhigste und beste Leben ist in Texas das Farmerleben und auch das billigste, denn das Vieh wächst hier und kostet kein Futter, wenig Mühe, zumal die Schweine, denn die vermehren sich in 1½ bis 2 Jahren von 6 bis 8 Mutterschweinen so sehr, daß für 4–500 Doll. können verkauft werden.

Nun muß ich Euch bemerken, dass es uns das erste Jahr durchaus nicht gefallen hat. Jetzt geht es etwas besser, aber der letzte Abend schwebt mir noch immer vor Augen, wie ich von Euch Abschied nahm, und nicht ohne Rührung kann ich mich erinnern. Im Traum habe ich oft mit Euch gesprochen, aber es war Traum, wenn ich erwachte. Mit uns gehts mit der englischen Sprache langsam. Wir können mehr verstehen als wir sprechen, und so muß Karl oder Christoph öfters für uns sprechen.

Aus Eurer Gegend haben wir wohl manches erfahren, auch selbst aus Ebersbach, durch das Löbauer Blatt, welches Bundesmann uns lesen läßt.

Wenn wir gleich 1800 Meilen geographisch von Euch entfernt sind, nähmlich die gerade Linie berechnet, denn der Tag differiert 9 Stunden, wenn Ihr Mittag esst, schlafe ich noch fest, früh 3 Uhr. Die Tageslänge ist jetzt früh von 6 bis 6 Uhr, im Sommer von 4–8; es ist auf einmal Tag und auch so Nacht, Dämmerung giebt es nicht. Wenn man eine Stange von 6 Fuβ senkrecht erstellt, so wirft sie einen Schatten 11 Zoll in den längsten Tagen. Verzeihet, daß ich so schlecht geschrieben habe; denn wir haben das Schreiben verlernt, und die Feder war auch nicht die beste. Noch eines muß ich bitten, schreibt mir mit, was macht mein Bruder, was machen seine Kinder? Freude darf ich wohl nicht hoffen, doch thut mir Bescheid von ihm, es sei schlecht oder gut.

Lieber Großvater!
Daß Ihr meine Handschrift auch noch seht, schreibe ich hier auch ein paar Zeilen und wünsche Euch viel Glück zum Neuen Jahr. Sollte der Brief auch noch nicht ankommen, so nehmet es dafür an. Ich bin bis jetzt noch gesund und diene in einem Bierhause in der Stadt, der Kirche gegenüber, die auf der Abbildung steht. Grüßet meinen Lehrer, den Herrn Kantor, von mir und saget Ihm, sein Liederkranz, welchen Er mir zum Abschied schenkte, mir viel Freude gemacht hat, zumal der Gang in die Heidelbeeren.
Ihr ältester Enkel Ch. F. Bergmann

Innigst geliebter Großvater und Pathe!
Ich kann nicht unterlassen, Euch ein paar Zeilen zu schreiben, daß Ihr seht, daß ich noch lebe, ich gehe jetzt in die Schule von früh bis Abend und muß viel lernen. Viele Grüße an meinen Lehrer und sagts Ihm, ich muß viel mehr lernen als in Deutschland.
Ich verbleibe
Euer zweiter Euch liebender Enkel
K.H. Bergmann

Werthester Großvater!
Daß ich noch lebe, seht Ihr an meinem Namen.
Christoph Wilhelm Bergmann

Unter vielen Grüβen verbleibe ich Euer
Euch liebender Sohn
Christian Friedrich Bergmann und
Joh. Christ. Luise Bergmann geb. Dressler
San Antonio, den 3ten Dezember 1856

[3.1]
San Antonio de Bexar, den 2. Juni 1857

Lieber Vater und Bruder, Schwestern und alle Freunde!
Ich fühle mich gedrungen, Euch wieder einmal einige Zeilen zu schreiben. Ich hätte Euch schon längst gerne geschrieben, aber ich wollte Euch doch Gewißheit von meinem Lande geben, wieviel und wo ichs habe. Auch war ich drei Wochen krank, auch Friedrich wurde zu gleicher Zeit krank mit einem Anfall von Scharlachfieber. Nachher wurde meine Frau krank, Anfall von kaltem Fieber, welche auch drei Wochen zugebracht hat, haben aber keinen Doktor gehabt.

Ich habe nämlich 320 Acker Land, in Komalkundi [Comal County] an der Quattlup [Guadalupe] lang gelegen, habe 2 Häuser darauf, aber noch wenig Fenß (Fence = Zaun), welches mich 200 Dollar kostet. Ich wollte dieses Frühjahr drauf ziehen, aber die große Trockenheit hielt mich davon ab, denn es hat vom halben September bis zum halben Mai nicht mehr geregnet. Den Osterabend regnete es, aber mit Hageln und nicht viel, worauf eine Kälte wurde, daß morgens 8 Uhr noch das Eis ziehmlich stark auf den Straßen war, welches noch nie vorgekommen war. Auch zuvor war schon zweimal das Korn erfroren, welches eine ziehmliche Teuerung verursachte; der Bushel (1 Bushel = 36,35 Liter) Korn (Mais) kostet ein Doll. und 80 Cent, und das ist wohl ziehmlich viel für schlechten Verdienst. Zum nächsten Herbst will ich aber doch auf das Land ziehen.

Friedrich und Karl sind diese Ostern konfirmiert worden. Friedrich ist noch in einem Bierhause und Karl ist in einem Laden, hat aber noch nicht viel Lohn, bloß 4 Doll. pro Monat, soll aber später mehr bekommen; ich will ihn aber auch nicht davon wegnehmen, denn er eignet sich doch nicht recht zur Ökonomie. Deshalb will ich ihn die Kaufmannschaft lernen lassen. Englisch lesen und schreiben kann er schon ziehmlich und so glaube ich, daß es ihm nicht schwer fällt zu lernen.

Auch muß ich bemerken, daß zwei Stunden von meinem Land ein Silberbergwerk gemacht wird, und [sie] haben auch schon Silber gefunden. In der Umgegend von und selbst in San Antonio ist die Dieb- und Mörderei sehr schlimm. Es ist in kurzer Zeit zu wiederholten Malen eingebrochen worden, auch Reisende ermordet und beraubt, Pferde, Maulthiere, Ochsen, Kühe und d.g.m. gestohlen worden. Aber jetzt scheint sich das Blatt gewendet zu haben. Letzten Freitag wurden drei von denen, welche man als Räuber und Spieler beschuldigt, vormittags erschossen. Mehrere wurden schon früher gehenkt und erschossen, und so will man jetzt immer noch fortfahren. Alle diejenigen, welche als Räuber und Mörder bekannt sind oder in Verdacht, sollen sich bis zum halben März gestellen. Widrigenfalls es nicht geschieht, werden sie, was ist einerlei, eben erschossen oder erhängt. Auch wurde ein Sheriff (Polizei) erschossen und ein Bürger schwer verwundet. Mehrere wurden verhaftet, welche man als schuldig erklärte.

Diesen Winter habe ich im Januar Schweine geschlachtet. Später habe ich Bäume veredelt, wo ich die letzte Zeit auch 2 Doll. verdiente pro Tag, welches ich auch wieder Anfang dieses Monats wieder unternehmen werde. Später bin ich 2 Meilen von der Stadt aufs Land gezogen, wo ich aber wegen der groβen Trockenheit, und wegen Hintrügungen, binnen 3 Wochen wieder in die Stadt gezogen [bin], wo ich sogleich wieder Arbeit bekam. Da habe ich auch geschwind das Maurerhandwerk gelernt. Es ist natürlich leichter als in Deutschland, denn die Steine werden entweder gesägt oder mit dem Beil zugearbeitet. Später will ich dann wieder Heu machen.

Lieber Vater und Geschwister, Euer Brief, welchen ich am 14. März durch Bundesmann erhalten habe, kam mir auch ganz unerwartet, aber desto mehr erfreulich, weil ich wieder auch Eure Handschrift zu lesen das Vergnügen hatte. Das Geld kam mir gerade zu gewünschter Zeit; ich hatte mir bei Bundesmann 125 Doll. gelehnt, wofür ich pro Monat 3 Doll. und 25 Cent Zins geben muβte, weil ich mein Heu noch nicht verkauft hatte und daher mit meinem Gelde nicht auslangte. Das Geld hat mir Bundesmann bar und richtig ausgezahlt, worüber ich quittiert habe und glaube, daβ auch Ihr die Quittung schon in der Hand habt. Meine Frau hat noch immer die gewöhnliche Arbeit, waschen und dann und wann etwas Nähen. Christoph geht in die Schule, vormittags und nachmittags, und hat sich seit einiger Zeit im Lesen und Schreiben so wie im Rechnen viel geübt.

Lieber Bruder, Du hattest mir geschrieben, daβ der Storch so viel Arbeit hat. Ist er denn nicht auch auf der Scheune gewesen? Denn in den Bauershöfen hält er sich auch gern auf. Schon zu meiner Zeit hatte er viel zu thun. Ich weiβ bis jetzt noch nichts, es muβ ihm vielleicht zu weit sein, und glaube nicht, dergleichen Vögel hier zu sehen.

Aber auch Dir, liebe Schwester, konnte ich nicht bei der Hochzeit beiwohnen, aber ich wünsche Dir von Herzen Glück zu Deiner Veränderung, daβ Du mit Deinem Mann ebenso in Frieden als im Glück miteinander leben möchte[s]t. Wollte Gott, daβ Euch mein Brief so gesund als mich der Eurige antreffen möge, und bitte zugleich, mir auch bald wieder einen zu schreiben. Grüβet mir eben alle Bekannte und Freunde. Gesund sind wir noch alle und wenn es jetzt noch einige Male noch regnet, so giebt es doch noch eine ziehmlich gute Ernte, und es wird dann auch bald wieder Leben in Texas

werden. Aber ich gehe ohne weiteres nächsten Dezember auf mein Land, denn es ist und bleibt das beste in Texas. Ich werde in kurzer Zeit wieder schreiben, denn es gehen zwei Reichenberger hier aus unserer Stadt wieder nach Deutschland, welche mir den Brief mitnehmen wollen.
Ich verbleibe Euer noch Euch liebender Sohn
Christian Friedrich Bergmann
San Antonio de Bexar, den 2. Juni 1857

[3.2]
[Texas, San Antonio de Bexar, den 12. Juni 1857]

Lieber Großvater!
Ich kann nicht unterlassen, Euch zu schreiben, mit im Namen meiner Brüder, daß Ihr daraus seht, daß wir immer an Euch denken und immer von Euch sprechen. Ich bin wieder gesund, denn ich lag schwer darnieder und dachte nicht, daß ich wieder gesund würde. In Ostern hab[en] ich und mein Bruder Karl das erste Mal das Hl. Abendmahl genossen, weil ein Evg. Luth. Prediger hier ist, mit Namen Zitzelmann, aus Württemberg. Ich diene jetzt in einem Bier-Hause und bekomme 7 Dollars pro Monat, weil die Zeiten zu schlecht sind, sonst bekäme ich 10 Doll.

Jetzt wollen wir aber noch in San Antonio bleiben, aber nächsten Herbst wollen wir auf unser Land ins Gebirge, 35 englische Meilen weiter nach Norden zu. Dort ist es gesünder als in der Stadt; denn ich hab voriges Jahr nicht weit davon schon Ochsen getrieben, da gefiel es mir sehr gut. In der Stadt ist jetzt viel Unruhe, denn es ist eine Räuberbande entdeckt worden, die ganz Texas unsicher machte, wo es öfters hieß, es wären Indianer, was nur wenig der Fall war; und jetzt fingen sie in der Stadt an zu rauben und mordeten unfern der Stadt 1 Farmer auf der Road (Straße). Nun hat sich eine geheime Gesellschaft gebildet, die ihn[en] nun nachgeht und wo sie einen von der Bande erwischen, gleich an einen Baum hängen oder erschiessen, wie es in der Mitte Mai der Fall war. Da wurden 3 auf der Straße erschossen, worunter ein Hauptmann von der Bande war. Den 11. dieses Monats haben sie 11 gehangen, 7 an einen Baum und einige Meilen vor der Stadt, dann 3 an einen Baum und einen, der alles gestanden hat, an einem Baum außer der Stadt. Da könnt Ihr Euch einen Begriff machen, wie es jetzt zugeht. Noch was muß ich bemerken, daß nur ein Deutscher unter der Bande ist, die anderen sind alles Amerikaner, Irländer und Mexikaner.

Viele Grüße an meine Freunde und Tischer Karl, sagt es, er soll so gut sein und grüßen alle meine Kameraden. Er kennt sie alle vollzählig,

Gochtliebels-August und den Herrn Kantor besonders. Und schreibt uns bald wieder, daß wir auch wissen, was bei Euch vorgeht. Verzeihet, daß ich so schlecht geschrieben habe, denn ich habe in der Eile des Abends geschrieben. Lebt wohl, denkt oft an uns!
Ich verbleibe
Euer Euch liebender Enkel
Christian Friedrich Bergmann
Texas, San Antonio de Bexar, den 12. Juni 1857

3. Farmleben an der Guadalupe in Texas: 1857 bis 1888

[4.]
Blanko County, den 16. Januar 1859

Lieber Vater und Geschwister und auch Schwager!
Lange, lange werdet ihr auf Antwort gewartet haben, aber ich habe erst den Brief im Dezember erhalten; denn wir waren lange nicht in San Antonio gewesen. Ich bin mehrere Male auf der Post gewesen, aber nie war ein Brief für mich da. Im Dezember kam ich wieder in die Stadt und ging zu einem Bekannten, der hatte den Brief auf der Post geholt, weil er in der Zeitung gestanden hatte, welcher mir denselben gab. Den nächsten Brief schickt Ihr nicht wieder nach San Antonio, sondern adressiert ihn wie unten steht. In Deinem Briefe, lieber Schwager, habe ich voller Neugierde gelesen und gefunden, daß ihr alle noch gesund und munter seid und daß auch die Luise geheiratet hat, wozu wir ihr alle von Herzen Glück wünschen. Und auch Dir, lieber Bruder, wünschen wir Glück in der Ekonomy.

Lieber Schwager, Du schreibst mir, daß es der Vater nicht mündlich von Bundesmann wollte glauben, wie wir uns befinden und so will ich jetzt anfangen. Ich habe 320 Acker Land, wovon ich jetzt zwischen 20 und 21 Acker in Fens habe, 700 Schritte habe ich Steinfens gemacht und das andere ist Holzfens. Das Land liegt an einem großen Fluß, die Quatelup [Guadalupe] genannt, und habe daher sehr gutes Wasser und auch, wenn wir Zeit haben, genug Fische. An Hochwild fehlt es auch nicht, nur hat es bis jetzt immer noch an Zeit gefehlt, welches zu schießen. Deshalb haben wir immer Fleisch zu essen, denn wenn ich einmal einen Hirsch schieße, so haben wir doch eine Zeit zu essen, oder einen Puter (Truthahn), die es hier in Masse giebt und auch eine vortreffliche Suppe giebt. Es giebt hier Puter von 24 bis 26 Pfund schwer.

Lieber Vater, ich will euch jetzt meinen sämtlichen Viehbestand schreiben: Ich habe 3 Joch Zugochsen, 6 Stück, 6 Kühe und 8 Stück Jungvieh, ein Pferd, vier Schweine, drey Hunde, wovon mich einer, fünf Wochen alt, 3 Dollars und einer, vier Wochen alt, 2 Dollar kostet; es sind dies englische Bull, welche gar keinen Spaß verstehen. Hühner habe ich noch nicht viel, gegen 20 Stück.

Was nun die diesjährige Ernte anbelangt, war bei mir nicht so glänzend, denn ich konnte zu spät pflanzen. Erstens war ich krank an der Ruhr, und zweitens krepierte mir ein Ochse, welchen wir lange gesucht haben, und so war jetzt guter Rath theuer, kein Geld und unbekannt dazu. Ich ging aber zu einem Amerikaner, drei Meilen von uns. Dort habe ich mir einen wilden Ochsen geborgt und in einem Monat bezahlt, welcher mich 18 Dollar kostet. Aber so Gott will, glaube ich dieses Jahr eine bessere Ernte zu machen. Zwei Acker habe ich Weizen gesät, welcher, wenn er einschlägt, 40 Bushel bringt und 3 Bushel sind so zirka ein Dresdner Scheffel, ein Bushel Weizen wiegt 60 Pfund. Auch will ich dieses Jahr Baumwolle und Zuckerrohr pflanzen. Diese beiden Artikel sind sehr einträglich hier. Freilich, die Baumwolle kostet viel Arbeit, aber wir sind Leute genug, sie zu besorgen. Auch habe ich große Zypressenbäume, von welchen die Schindeln gemacht werden. Wenn ich sonst keine Arbeit habe, so mache ich Schindeln, denn diese sind hier leichter zu machen als in Deutschland, weil sie nicht genutet werden, sie werden gedeckt wie bei euch die Ziegeln, und nach dem Tausend verkauft werden, welche, wenn sie einen guten Preis haben, für vier und fünf Dollar verkauft werden. Und so habe ich doch immer ein gutes Tagelohn. Wenn ich jetzt mit dem Pflügen fertig bin, so muß Friedrich nach San Antonio fahren mit Zedernholz für Fens, wo, wenn das Holz einen guten Preis hat, für 100 Stück ich 8 bis 10 Dollar bekomme, welches mir dann wieder Kaffee, Zucker, und andere Kleinigkeiten ins Haus bringt.

Lieber Schwager, Du schriebst mir, daß Du schon oft von uns geträumt hast. Dasselbe ist auch bei mir und bei uns allen der Fall gewesen, aber allemal habe ich mich geärgert, daß ich mein Grundstück verlassen habe. Ich werde höchstwahrscheinlich nicht wieder nach Deutschland kommen, denn so weit kann ich es nicht in Deutschland bringen, wie ich es hier habe. Das Vieh wächst mir heran und sobald ich kann, will ich auch noch mehr Kühe kaufen. Denn es ist ganz einerlei, ob ich 10 Stück Kühe oder 30 habe, ich muß doch wöchentlich zweimal nach dem Vieh sehen. Ich kann wohl sagen, daß mir mein Vieh wenig zu schaffen macht; denn allemal, wenn es zu Hause kommt, giebt ihm meine Frau Salz zu lecken, und wenn das regelmäßig der Fall ist, so geht das Vieh nicht weit von der Farm weg. Übrigens habe ich, Gott sei Dank, noch keinen Verlust an Vieh gehabt.

Auch habe ich nicht weit in die Mühle, bei welcher zugleich auch eine Sägemühle ist, wo ich mir verschiedenes Holz zu Brettern und Pfosten kann schneiden lassen, und nach der Stadt bringen kann. Von meinem ganzen Vermögen gebe ich jährlich 4 Dollar und ½ ab, wobei auch das kleinste Haus mehr Abgaben hat. Es ist freilich wahr, daß wer nichts gelernt hat, nur nach Texas gehen soll. Ich bin hier alles, Schuster, Stellmacher, Kaltschmied, Fleischer, kurz, alles, was vorkommt, wird gemacht. Auch die Ochsen und Schweine schneide ich mir selbst, und sogar den Nachbarn muß ich sie schneiden.

Was nun das übrige anbelangt, von wilden Früchten haben wir Kirschen, Pflaumen, Wein, frühen und späten. Ich habe dieses Jahr noch keinen Wein gemacht, aber nächstes Jahr will ich auch welchen machen. Es giebt Farmer, welche 10 bis 14 Eimer Wein gemacht haben. Ich habe es aber noch nicht so getraut. Pfirsichbäume habe ich gegen 1500 Stück, 50 davon sind ganz große Sorten. Die Pfirsichbäume tragen schon das dritte Jahr vom Kern aus hier in Texas. Auch habe ich auch einige grüne Pflaumenbäume und Aprikosen, und nun wünschte ich mir nur, daß ich von meinem Kirschensortiment einige Bäumchen hätte. Apfel- und Birnbaum kann ich hier bekommen.

Ich will jetzt meinen Brief schliessen, wir sind alle gesund und wünschen, daß auch dieser Brief Euch gesund antreffen möge, und bitte um recht baldige Antwort. Freilich, ich hätte wohl einen Monat früher schreiben können, aber es war allemal etwas, was mich abhielt, und es hat sich so weit herausgezogen. Bitte daher alle, es mir nicht übelzunehmen. Meine Adresse ist Mr. C.F. Bergmann, Hodges Mill, Blanko County, Texas.
Ich verbleibe, Euer [Euch] liebender Sohn
Christian Friedrich Bergmann

Lieber Bruder, grüße mir alle Bekannten und Freunde, und besonders vorzüglich Wendebaum, Stellmacher Weber und Bergmann auf der Haine gieb den Brief zu lesen, denn sie haben mich gebeten, an sie zu schreiben. Meine Blumenkollegen grüße mir ebenfalls, und ich wünsche ihnen allen Glück und Gesundheit, und die Blumen habe ich vergessen. Dann und wann zuckt es noch ein wenig nach.

Lieber Vater, schickt Euren Brief über Bremen und New York.

[5.]
[Blanco County, den 1. Dezember 1859]

Lieber Vater!
Euren Brief habe ich erhalten und daraus gesehen, daß Ihr noch alle gesund seid. Lieber Vater, Ihr meint in Eurem Briefe, daß ich Euch zu wenig geschrieben habe von meinem Lande; ich hatte wohl so viel in Fens, aber noch nicht so viel gebrochen, ich hatte bloß 15 Acker gebrochen. Davon habe ich 9 Acker mit Maiskorn und 3 Acker mit Baumwolle gepflanzt. Auf die 2 Acker, auf welchen ich den Weizen hatte, habe ich, nachdem ich den Weizen rein hatte, den 4. Juli noch Mais gepflanzt, in der Meinung, Futter für die Ochsen zu machen. Aber das Wetter war günstig, da habe ich noch schönes Korn bekommen (Mais nennt man Korn). Auch habe ich für 29 Dollar Melonen verkauft, welche ich in die Baumwolle gepflanzt hatte. Es waren vierhundert Stück, ich habe aber viel mehr gezogen. Sie hatten aber keinen Preis mehr in San Antonio, und so habe ich sie mit den Schweinen gefüttert, wo ich jetzt 27 Stück habe. Drei davon will ich schlachten und auch einen alten Zugochsen, denn er springt mir immer über die Fens und hat mir viel Schaden gemacht. Denn es ist nicht genug, daß er allein hereingeht, sondern das andere Vieh geht nach und gewöhnt sich auch daran.

Meine diesjährige Ernte war bedeutend besser als die vorjährige. Ich glaube, daß ich zwei Ballen Baumwolle ernte und der Ballen zu 500 Pfund. Und der Ballen kostet 50 Dollar. 120 Bushel habe ich Korn geerntet. Davon will ich 70 Bushel verkaufen. Noch habe ich zu bemerken, daß ich auch chinesisches Zuckerrohr gepflanzt habe, aus welchem der Sirup gemacht wird. Ich habe bloß einen Versuch gemacht für dieses Jahr, aber nächstes Jahr will ich mehr pflanzen. Ich habe doch gegen 80 Kannen Sirup gemacht. Das Zuckerrohr wird 8 bis 10 Fuß lang und bringt viel Samen, welchen ich mit den Schweinen füttere. Ich habe mir auch die Presse selber gebaut, welche mit zwei stehenden Walzen von einem Pferde gedreht wird. Das hat ein schweres Stück Arbeit für mich gegeben, denn ich konnte ja die Walzen nicht abdrehen, sondern ich mußte sie bloß hobeln. Das hat mir viel Mühe gekostet, um sie rund zu bringen. Auch habe ich 450 Kürbisse geerntet, wovon wir die schönsten selbst essen und verkaufen; die anderen füttere ich eben mit den Schweinen, welche gekocht eine gute Mast geben. Für dieses Jahr habe ich auch einen kleinen Versuch mit süßen Kartoffeln gemacht, aus dem Grunde, weil ich selber keine Pflanzen hatte, und es waren in der That auch wenig zu bekommen, aber ich habe auch sehr schöne gezogen. Wieviel, das weiß ich noch nicht, denn ich habe sie noch nicht ausgemacht. Denn das Baumwollelesen hält mich zurück.

Dieses Jahr habe ich 2 Acker Winterweizen gesät, 2 Acker will ich Roggen und zwei Acker Sommerweizen säen. Das Dreschen des Weizens hat mich getrubelt (trouble = Mühe, "getrubelt" = Mühe bereiten), keinen Flegel, keine Scheune, ich habe bloß im Hause müssen dreschen. Die Flegelkappen und Mittelbänder habe ich von Rohhaut gemacht, Flegel und Handhaben habe ich gehobelt, das Handhaben-Machen hat mir am meisten zu schaffen gemacht, aber ich habe sie auch fertiggebracht. Auch will ich einen halben Acker mit Tabak pflanzen für dieses Jahr. Ich würde wohl mehr pflanzen, aber ich habe noch zu wenig Erfahrung in der Behandlung derselben. Mein Nachbar hat 1½ Acker gepflanzt und hat über 2000 Pfund geerntet. Er hatte aber einen Mann, der die Sache verstand, und der Ertrag dieser Ernte kommt über 250 Dollar.

Ich bin jetzt an meinem Haus bauen, welches ich von Zederholz baue. Das Zederholz ist dasselbe Holz, wie in Friedersdorf der Gartenzaun ist. Das Haus wird 48 Fuß lang und mit Halle und vorn mit Galerie, 36 Fuß breit und in der Mitte einen Durchgang mit 12 Fuß breit. Die Bauart des Hauses ist, man legt die Bäume übereinander und klinkt sie an beiden Enden ein, so daß sie nicht aufeinander können. Der Zwischenraum wird mit Steinen und Lehm ausgeschlagen und die Bäume werden inwendig und auswendig behauen. Wenn ich das Haus fertig habe, dann kann ich im Durchgange dreschen.

Lieber Vater, Ihr wolltet aber auch wissen, was wir mit der Butter und der Milch machen. Das ist nicht so wie in Deutschland, daß eine Kuh 9 bis 10 Pfund Butter giebt. Die Kälber saugen, bis sie Dreivierteljahr alt sind und oft noch länger, je nachdem die Kühe wieder Kälber machen, und man darf da nur die halbe Milch nehmen, denn die Kälber dürfen den Tag über nicht aus der Panna (Panna ist ein mit Stoffetzen eingezäuntes Stück). Für die Kühe ist Salz das einzige, was man ihnen giebt, und sie wollen auch nichts anderes, denn sie sind es nicht gewöhnt, und so habe ich noch keine Bange, wo ich mit Butter und Milch hin soll. Wenn wir genug haben, so bin ich froh. Denn wenn es die Mutter sähe, wie wir Butter essen, so würde sie sprechen: "Ihr klötzelt", und die Nachbarn sprechen deshalb immer, wir sind genau. Aber unser Klima verlangt, gut und fett zu essen, sonst würde die Arbeit sehr hart werden.

Auch muß ich bemerken, daß eine Krankheit unter Rindvieh und Schweinen war, sogar unter das Hochwild kam sie. Auch ich verlor durch sie 3 Säue und einen jährigen Ochsen, und ein Kalb wurde mir blind. Es thut aber hier nicht so wehe wie in Deutschland.

Lieber Schwager, Du schreibst mir, daß Freude gern von Heuchling wollte wissen, was er machte. Ich habe mit Kastroviller Leuten unterwegs nach San

Antonio gesprochen, welche sagten, er habe anfangs schönes Geld verdient mit dem Musikmachen, das habe aber jetzt aufgehört, und dabei habe er sich von der Arbeit etwas abgewöhnt Auch habe ich mit Feines Bruder Tochter gesprochen, welche mir sagte (sie diente eben beim Gärtner Marx in San Antonio), daß er mit seinem Bruder in stetem Unfrieden lebte. Ich meinerseits habe noch nicht mit ihm gesprochen. Er ist in San Antonio gewesen, aber er ist nicht zu mir gekommen; so viel weiß ich aber, daß es nicht am besten mit ihm steht.

Lieber Vater und Bruder, ich muß Euch noch mit einem Auftrage belästigen, ich bin gesonnen, mir einen neuen Wagen bauen zu lassen und zwar bey Gottlieb Weber und Ulrich in Ebersbach. Ich muß ihn aber so stark haben, daß ich kann 60 Zentner aufladen und über Stock und Stein fahre[n]. Ich will aber bloß die vier Räder und die zwei Schenkel, aber eiserne Schenkel. Die Schenkel dürfen aber nicht mehr als zwei Zoll vorstehen, denn die Wege sind hierzulande stellenweise sehr eng wegen den Bäumen. Die Räder dürfen auch keinen Sturz haben, da hier kein Spaßen ist. Denn bei schlechtem Wege gehen die Wagen manchmal so tief im Geleise, daß die Achsen auf dem Boden aufliegen. Aber ich will auch keine hellweißen Speichen in den Rädern haben, überhaupt gutes Holz. Deshalb soll mir Weber den Wagen bauen. Die vorderen Räder müssen hoch sein 2 Ellen, 7 Zoll und die hintern Räder 2 Ellen, 19 Zoll. Lieber Vater, bestellt aber den Wagen sogleich nach Empfang des Briefes, daß das Holz gut trocknet, so daß ich den Wagen nächsten Herbst bekomme. Das Geld für den Stellmacher hoffe ich, werdet Ihr wohl auslegen. Denn ich habe noch nichts von meiner Ernte verkauft. Über die Bezahlung, lieber Bruder, sprich nur mit Bundesmann. Der weiß Dir Bescheid zu geben, wie ich es am besten kann abmachen, ob in Wechseln oder Geld. Schreibt mir aber sobald als möglich, damit ich weiß, wie teuer der Wagen kommt, oder ob Ihr Euch das nicht unterziehen wollt. Noch habe ich vergessen, wie breit die Spur des Wagens sein muß, sie ist 5 Fuß, 2 Zoll breit im Licht, und keine Deichsel und keinen Scherhaken. Denn in einem Ochsenwagen muß die Deichsel sich von oben nach unten bewegen; das will ich mir selber machen. Aber einen zweckmäßigen Schraubenschüssel wünschte ich mir noch, damit ich auch kleinere und größere Schrauben kann anschrauben.

Ich habe jetzt vier Joch Ochsen, ich habe einen an die Stelle des verendeten gekauft, für 23 Dollar, und einen jungen Ochsen zu meinem, den ich selbst aufgezogen habe. Den habe ich billiger, für 8 Doll.; und ich hoffe, daß ich im nächsten Herbst 5 Joch Ochsen habe. Denn ich habe wieder einen jungen selbst, und einen will ich wieder dazu kaufen. Damit will ich meinen

Brief schließen und wünsche, daß er Euch gesund antreffen möge. Ich verbleibe
Euer Euch liebender Sohn, Bruder und Schwager
Christian Friedrich Bergmann
Blanco County, den 1. Dezember 1859

[6.1]
[San Antonio, 3. März, kein Jahr, wahrscheinlich 1861]

Lieber Vater und Geschwister!
Ich habe den Brief von Richtern erhalten und habe mich sehr gefreut, daß Ihr alle noch gesund seid. Ich habe den Brief aber erst den 3ten Februar erhalten, da wir seit Weihnachten nicht mehr nach San Antonio gekommen waren und Richter ihn nicht auf die Post gegeben hatte. Erst wie ich zu ihm kam, gab er mir selbigen. Richter hatte eine schöne Reise gehabt und ist gesund und munter in San Antonio angekommen mit seinem Vater und Schwester.

Lieber Vater und Geschwister, wir haben eine sehr bedenkliche Zeit in Amerika, die südlichen Staaten wollen sich von den nördlichen trennen, und es haben sich auch wirklich schon 8 derselben vom Norden getrennt. Die Trennung kommt lediglich von dem neuen Gouverneur her, denn es wurde im Herbst ein neuer gewählt. Der Norden ist gegen die Sklaven, und der Süden ist für dieselben, und so gelang es dem Norden, den Gouverneur gegen die Neger zu wählen. Darüber wurde der Süden empört, weil da viele Sklavenhalter sind, und infolge des kommt es zur Trennung. Dieser Umstand giebt Texas einen ungeheuern Schlag. Es schlägt sich natürlich auch zum Süden, und in kurzer Zeit wird das Gouvernement von Texas weggenommen werden und zu gleicher Zeit auch die Soldaten, welche an der Grenze der Indianer stehen. Es hat schon jetzt einen bedeutenden Eindruck auf San Antonio gemacht. Alle Rohprodukte haben bedeutend an Preisen verloren.

Lieber Vater, daß mir Richter den Wagen nicht mitgebracht hat, ist mir ein großer Schaden, denn ich hatte viel zu fahren. Mein alter ist sehr schwach und auch sehr ruiniert, und wenn ich ihn sollte reparieren lassen, so würde er auch beinahe so viel wie ein neuer kosten, und so kann ich jetzt wenig und gar nichts fahren. Ich muß alle meine Ochsen nur so laufen lassen. Freilich habe ich mein Feld gut in Ordnung gebracht und habe auch schon wieder auf dem neuen Land eingefenst. Denn Friedrich und Karl sind jetzt schon stark genug, mir bey solcher Arbeit zu helfen. Denn Friedrich ist beinahe so stark wie ich und Karl nicht viel weniger. Christoph ist bei einem Nachbar und treibt da

Ochsen und thut andere Arbeit, wobei er 5 Doll. monatlich verdient und die Leute sehr zufrieden sind.

Lieber Schwager, Du fragst, ob sich nicht auch Störche sehen liessen, in Deinem Briefe. Wir kennen derartige Vögel nicht hier in Texas, und ich glaube nicht, solche mehr zu sehen zu bekommen. Wir werden alt und bekümmern uns nicht mehr viel um sie.

Lieber Vater, ich bedauere nur, daß ich mich eingelassen habe, einen zweiten Wagen bestellt zu haben, denn mein Nachbar Wenzel hat noch nicht einmal nach demselben gefragt, und ich bin auch selbst zerfallen mit ihm. Ich habe nähmlich fünf Acker freies Land eingefenst, welches gerade an meine Fens anstieß. Dieses Land wollte ich später kaufen, aber Wenzel ging zum Vermesser und ließ es sich vermessen, ohne daß ich etwas wußte. Dieses hat mich freilich sehr geärgert, und ich bin seit der Zeit nicht wieder in sein Haus gekommen, er aber auch nicht mehr zu mir. Jedoch schickt den Wagen, denn es will ihn ein anderer Nachbar haben.

Lieber Schwager, Du hast mir aber gar nicht von meinen alten Kameraden und Blumästen geschrieben, ich möchte gerne wissen, was sie machen. Auch an meinen Gevatter Bergmann auf der Haine sage viele Grüße und daß wir noch alle gesund wären, und ich könnte mich nur noch wenig an die Blumen erinnern, ich hätte sie alle nach und nach vergessen, aber dass ich einen ziehmlichen Baumgarten angelegt habe von Äpfeln, Birnen, Pflaumen, Kirschen. Pfirsch, usw. Den Samen habe ich mit den Marunkenkörnen auch erhalten, und ich habe mich sehr gefreut.

Die Schenkel werdet Ihr wohl wahrscheinlich in eine Kiste packen, da seid so gut und nehmt zur Eingelage etwas Werg dazu. Denn wir brauchen es sehr nothwendig zum Gewehr ausputzen, denn hier können wir keins bekommen. Hier schicke ich Euch einen Wechsel von 64 Thaler 28 ngl. Es thut mir leid, daß ich einen Wagen [nicht] ganz bezahlen kann, aber ich brachte das Geld nicht ganz zusammen. Ich mußte 3 Kühe an einer Schuld nehmen, aber ich glaube, daß ich, wenn ich die Rechnung erhalte, daß ich es dann bezahlen kann. Versicher die Wagen und was die Fracht bis nach Bremen oder Hamburg macht, bringt Ihr mit mir zur Rechnung. Die Wagen schickt an A. Runge und Co. Indianola, Texas.

Lebet wohl und ich wünsche, daß Euch dieser Brief gesund in Eure Hände bekommt. Wir sind alle gesund und grüßen Euch alle.

Euer [Euch] liebender Sohn—Christian Friedrich Bergmann

Blanco County, Hodges Mills

Hier inliegend einen Dollar für den Stellmacher Weber, an Werth 1 Thaler 1 Onze

[6.2]
[Loser Zettel bei den Originalbriefen; ohne Datum]

Lieber Vater,
wenn Ihr mein, das bei Euch stehende Geld noch nicht angesandt oder an H. Bundesmann in Neugersdorf ausgezahlt habt, so bitte ich Euch, 150 Thaler Prauschkurant an selben umgehend zu entrichten und das Übrige in einem Wechsel auf New York durch ihn mir zu senden oder das ganze Geld gegen eine Anweisung an E. Bundesmann hier in San Antonio an Herrn Bundesmann in Neugersdorf zu zahlen. Sobald ich die Anweisung habe, bekomme ich das Geld von seinem Sohn hier ausgezahlt.

Es liegt mir doppelt daran, daß Ihr diese Angelegenheit sobald wie möglich regelt, indem ich schon Geld darauf empfange.
Euer [Euch] liebender Sohn
Christian Friedrich Bergmann
San Antonio, 3. März

[7.]
Guadelup, den 1. Juli 1865

Lieber Vater, Bruder, usw.!
Endlich ist es mir wieder vergönnt, an Euch zu schreiben. Es sind nun 4 Jahre her, daß ich nichts mehr von Euch gehört habe. Aus dem letzten Brief habe ich ersehen, daß Ihr noch alle gesund wart, worüber ich mich sehr gefreut habe. Lieber Bruder, Du schriebst mir wegen der Wagen, ob Du sie über New York schicken solltest, aber wie ich den Brief bekam, war schon der Hafen blockiert, und ich konnte Dir keine Antwort schreiben. Die Rechnung von diesen Wagen habe ich erhalten und sie ist mir auch nicht zu hoch. Ich habe einen ziehmlich starken Wagen gekauft, auch mit eichenen Achsen, auf welchen ich auch 60 Zentner aufladen kann und 330 Dollar bezahlt habe. Es hat mir freilich viel geschadet, daß die Wagen nicht kommen konnten, denn es war reichlich Geld zu verdienen. Ich glaube aber und bin in der Hoffnung, daß ich die Wagen doch noch bekommen kann, und ich brauche sie auch, wenn auch nicht gleich. Doch kommt die Zeit, daß Friedrich und Karl auch Wagen haben wollen und ich die beiden für sie bestimmt habe. Ich würde den Rest für die Wagen mit in den Brief gethan haben, aber Posten gehen noch zu unsicher, [als] daß man

kann Wechsel aufgeben. Schreib mir, lieber Bruder, wie es mit den Wagen steht und, sollte eher ein Schiff abgehen, als ich wieder schreiben kann, so schicke sie und zwar wieder an Herrn Runge & Compagnie in Powderhorn. Sobald ich wieder Antwort von Dir habe, so will ich das Geld für die Wagen schicken und womöglich auch den verlangten Golddollar. Ich habe zwar schon einen Brief mit 3 Dollar Gold herausgeschickt, aber der Brief muß verloren sein.

Lieber Bruder, Du hast mir Deinen sämtlichen Viehstaat geschrieben, und Du wolltest aber auch gerne meinen Viehbestand wissen: Ich habe erstens 3 Reitpferde, ein Maulthier, auch zum Reiten, 2 alte Stuten, 2 zweijährige Stuten, 2 einjährige Stuten, 2 zweijährige Wallache, 4 jährige Hengste und ein Fohlen, 16 Zugochsen, 31 Kühe, 11 wilde Ochsen 3 Jahre alt, 14 Kalben, 24 einjähriges Vieh, 16 Kälber, also weniger als im vergangenen Jahre. 12 Kühe sind krepiert (es war eine Krankheit unter dem Vieh), und die anderen machen keine Kälber. Schweine wenig, 19 Stück; Schafe 44 Stück, welche aber nicht mein sind, sondern Karl und Christoph gehören; 30 Hühner, 3 Paar Tauben, 7 Gänse und 20 Enten. Das ist mein sämtlicher Viehbestand.

Von dem Politischen kann ich Euch nicht viel schreiben, wollte ich das, so würde ich nicht fertig. Die Jungen sind alle drei mit einem blauen Auge davongekommen, Friedrich hatte die Nervenkrankheit, hat aber doch eine Zeitlang fürs Gouvernment fahren müssen, Karl war immer nicht alt genug, sie sind aber sehr gewachsen. Friedrich ist so groß wie ich selber, Karl ist ein und ein halb Zoll größer und Christoph ist auch bald so groß wie ich, genug, wir können alle einen Schuh brauchen. Übrigens sind wir alle gesund und wünschen, daß Euch diese Zeilen gesund und munter antreffen. Herzliche Grüße von uns allen
Euer Sohn und Bruder
Christian Friedrich Bergmann

Adresse: C.F. Bergmann
Boerne, Kendall County, Texas

[Lose beiliegender Zettel zu diesem Brief]
Diese Zeilen wünsche ich wieder retour. Lieber Vatter, ich habe von Bundesmann 390 Dollar richtig erhalten.
Christian Friedrich Bergmann

[8.1]
Guadelupe, den 28. Juni 1866

Es war Montag, den 25. Juni. Wir hatten eben unsere County und Staatswahl in Boerne, als ich in Begleitung mit Friedrich und seinem Schwiegervater zur Wahl ritten, um unsere Stimme abzugeben. Zu gleicher Zeit wollten wir aber auch die Zeitung mitbringen, denn es war gerade Posttag, und so verweilten wir noch ein paar Stunden, bis die Post kam. Ich schickte dann gleich Friedrich nach den Zeitungen, und zu meiner größten Freude brachte er auch den Brief von Euch mit, welchen ich folglich aufbrach und las, und zu meiner größten Freude habe ich gelesen, daß Ihr alle noch gesund seid; bis aufs Große und seine Frau und die Christliese. Jetzt entdeckte aber Friedrich die drey Bilder von Euch. lieber Vater, welche bald durch viele Hände gingen. Der Kaufmann, bey welchem wir waren, schnitt sogleich drey Gläser, denn er sagte, das wäre ihm ein zu ehrlich Gesicht, da müßte er die Gläser zum Einrahmen geben. Nachher haben wir noch einige Gläser Bier auf Eure Gesundheit getrunken und sind dann im Galopp nachhause geritten, wo wir unsere Leute schlafend antrafen, denn es sind 15 Meilen. Aber wie sie hörten, daß ich einen Brief von Deutschland hatte, waren sie alle wie eine aufgescheuchte Schaar Vogel munter. Nun könnt Ihr Euch wohl denken, daß es allen eine herzliche Freude war. Meine Frau, lieber Vater, hat den Brief mit einer Träne im Auge gelesen, und es haben sich auch alle herzlich gefreut, daß Ihr noch lebt und gesund seid, und auch über den bildlichen Großvater.

Lieber Vater und Geschwister usw., meine Wirtschaft hat sich aber auch wesentlich verändert. Friedrich hat sich verheirathet mit einer Nachbarsfarmer Tochter, welche aus Oberschlesien hierher gekommen sind. Karl ist auch verlobt, aber will doch noch ein Jahr warten, ehe er heirathet. Dieses Mädchen ist von einem Deutschen soviel als Pflegemädchen angenommen, und ihre Mutter macht ihm die Wirthschaft. Er ist unverheirathet, welches aber die Haupttriebfeder zum Heirathen ist, weil er selber nicht heirathet und sich auch nicht mehr allein so plagen will. Denn er sagte mir, er wüßte nicht, wie lange er noch lebe, und so soll Karl die Farm übernehmen.

Lieber Vater, Ihr wolltet gerne wissen, was ich für ein Haus habe, und so schicke ich Euch den Grundriß mit. Ich habe einen Freund, welcher sehr gut zeichnen kann. Wenn er einmal zu mir kommt, so will ich ihn das Haus lassen abzeichnen und es Euch schicken. Ich versichere Euch, es ist das schönste und beste weit und breit, aber es kostet mich auch viel Arbeit und Geldt. Ich werde aber auch von manchem beneidet darüber. Es war neulich einer da, welchen meine Frau nötigte, daß er ein ander mal seine Frau sollte mitbringen. Er sagte aber, das dürfte er nicht thun, sonst würde er zuviel gequält, wenn sie jetzt unser Haus sieht.

Lieber Vater, Ihr wollt auch gerne wissen, was für Holz auf meinem Lande wächst. Es sind vier verschiedene Sorten Eichen: die Lebenseichen, Pooβoak-Eichen, Schwarzweichen und Weiβeichen, aber Ihr dürft Euch nicht vorstellen, daβ diese so stark werden wie in Deutschland. Auch habe ich noch andere Sorten Holz an der Guadelupe, welche ich für Nutzholz gebrauche: die Rothulme, die Schlickerulme und die Platane. Auch habe ich noch Ziebressen am Wasser, welche wir für Schindeln verarbeiten, von einer beträchtlichen Stärke. Ich habe dieses Frühjahr eine umgehauen von fünf Fuβ Durchmaβ, und mein Nachbar eine von sieben Fuβ. Auf dem anderen Stück Land ist hundert Acker Zederwald (Duja), welches ich nach San Antonio fahre und auch sehr gut für Fensriegel ist, und achtundzwanzig Acker wieder mit verschiedenen Sorten Nutzholz, Bockannuβ, Schwarzwallnuβ und Häckbärbaum, auch etwas Linden; Kiefern, Fichten, Lärchen, Erlen und Birken sind hier in West-Texas keine, in Ost-Texas sind Kiefern. Noch muβ ich bemerken, wie groβ ein Acker ist, er ist 75 Schritte lang und ebenso breit, und so habe ich 608 Acker, aber in zwey Stücken. Der Zederwald ist 3 Meilen von meinem Hause.

Ihr wollt auch wissen, was überhaupt auf meinem Land wächst. Es gedeihen alle Früchte gut, wenn die Witterung günstig ist, aber zuweilen bleibt der Regen aus, und so ist es manchmal verfehlt. Es liegt überhaupt viel daran, wie man es mit den Pflanzen trifft. Meine diesjährige Ernte wird noch besser, als ich erwartet habe. Ich habe dieses Jahr als Haupternte nur Mais, Weizen und Roggen habe ich keinen dieses Jahr, Weizen und Roggen; Erbsen, Zuckerrohr für Syrup, süβe Kartoffeln und Millet. Millet ist eine Sorte Gras, welches man im Winter für die Pferde braucht. Dies gedeiht alles gut. Was die Obstbäume betrifft, so sind die Pfirsichbäume vorzuziehen. Äpfel und Birnen gedeihen wohl auch. Ich habe vergangenes Frühjahr ziemlich viel Pfirsichbäume gepflanzt. Sie tragen gewöhnlich schon das zweite Jahr, nachdem man sie geplanzt hat und wachsen fabelhaft schnell. Von Gemüse kann man alles pflanzen, ich habe vergangenen Herbst Salat gesät, wovon wir bis April davon gegessen haben. Ich habe einen schönen Garten für Gemüse und Bäume und habe auch eine Quelle in demselben, aber nicht zum Schöpfen, sondern ich muβ das Wasser eine kleine Strecke in Rinnen leiten, aus welchen ich es in einem groβen Troge auffange, welcher ungefähr 10 Tonnen hält, welchen ich aus einer Zipresse gemacht habe. Nächsten Herbst will ich auch wieder Weizen und Roggen säen.

Ich habe auch seit meinem letzten Briefe ziemlich Verluste gehabt, nämlich 3 Paar der besten Zugochsen und 3 Kühe, aber der Schaden ist wieder geheilt. Was die Pferde anbelangt, so habe ich besonderes Glück gehabt. Ich habe 4 Pferde zum Reiten und ein Maulthier, wovon ich eins Friedrich gegeben habe. Übrigens hat Friedrich 2, Karl 2, und Christoph 1 Pferd. Übrigens habe ich noch 11 Stück, welche zur Zucht bestimmt sind, wobei auch 2 zahme Stuthen sind. Jetzt brauchte ich einen guten Hengst zu halten, dafür sind es zu wenig Pferde, denn ein guter Hengst kostet 4–500 Dollar.

Lieber Bruder, Du schreibst in Deinem Briefe, daβ ich so wenig geschrieben habe. Das ist wohl wahr, aber wenn ich an die Vergangenheit denke, so ekelt's mich an, überhaupt, ich würde nicht fertig. Wir haben uns so gut wie möglich durchgeschlagen, dass wir nicht durften Soldat werden. Es sind in den letzten verlebten Jahren so viel Greuelthaten verübt worden, wie sie wohl keine der französischen Revolutionen kann aufweisen. Die von der Gegenpartei haben mehrere Male angesetzt, uns zu Soldaten zu machen, aber es war vergebens. Zum Abfangen ging es nicht, und gutwillig gingen wir auch nicht. Mein Haus ist überhaupt so gut gelegen zum Entlaufen. Ich habe das letzte Jahr im Kriege viel Besuch gehabt, nämlich Deserteure. Woltersdorf aus San Antonio ist ein volles Jahr bei mir mit seiner Familie gewesen. Woltersdorf ist der Schwiegersohn von dem alten Richter aus Groβhennersdorf. Woltersdorf will nächsten Herbst wieder nach Löbau kommen, aber wie es jetzt bei Euch aussieht, wird auch nichts werden. Er hat mir versprochen, wenn er wird drauβen sein, daβ er zu Euch kommt, und dann hört Ihr mündlich. wie es um mich steht.

Lieber Vater, ich habe mir vorgenommen, mit Hermann Woltersdorf die Sache abzumachen, da den Weg, wenn er nach Deutschland kommt, die wenigsten Schwierigkeiten sind. Ich wünsche, daβ Euch der Brief gesund antreffen möge. Ich verbleibe
Euer Euch liebender Sohn und Tochter und Enkel
Christian Friedrich Bergmann
Johanna Christiane Luise Bergmann
[Dieser Brief enthielt eine Zeichnung von Bergmanns Hausgrundriss.]

[8.2]
[Dieser Brief wurde aller Wahrscheinlichkeit nach zusammen mit dem vorherigen Brief mit dem Datum "Guadelupe, den 28. Juni 1866" geschickt, da das gleiche, von Bergmann selten gebrauchte blaue Briefpapier benutzt wurde und auch das Schriftbild der beiden Briefe einheitlich ist.]

Lieber Bruder, Du willst gerne wissen, ob Du nach Texas kommen sollst oder nicht. Das ist eine schwere Aufgabe für mich. Ich habe schon von vielen gehört, dass sie sich den Weg viel Vorwurf zugezogen haben, und das wollte ich doch nicht haben. Denn ich nehme es an uns aus, wie wir ins Land kamen. Hätte da jemand die Schuld gehabt, ich würde ihm die bittersten Vorwürfe gemacht haben, und somit rathe ich Dir, nicht zu kommen. Aber ich sage auch nicht, daß Du sollst draußen bleiben und Dich Dein Leben lang in Schulden rum quälen sollst. Es ist mit Dir eine ganz andere Sache wie bey mir. Ich kann Dir mit Rath und That helfen, was bey mir nicht der Fall war, und Du findest auch hier Freunde. Alles war bey mir nicht. Das gesellige Leben ist hier auf dem Lande nicht so wie in Deutschland. Die Häuser sind zu weit voneinander, dass Du mit den Nachbarn sprechen kannst, und so entwickelt sich bald das Heimweh. Willst Du dieses nicht so hoch anschlagen, so will ich sagen, komm nach Texas. Es ist noch genug gutes Land zu kaufen. Freilich wirst Du Dich auch nicht für den ersten Anblick sehr an Texas ergötzen, wie mir es auch ging. Aber wenn es der gute ernste Wille ist, so gewöhnet sich das Auge bald daran. Willst Du kommen, so ist ein gutes Stück Land 3 Meilen von mir zu verkaufen. Der Mann war kürzlich bey mir, und ich habe ihm gesagt, daß Du Lust nach Texas hast, und so hat er mir versichert, daß er das Land nicht eher will verkaufen, als bis er weiß, daß Du nicht kommst. Du kannst aber auch ein oder zwei Jahr renten und dabei Dir ein Stück Land zurecht machen. Dabei kannst Du Dir, wenn Du gesund bist, noch etwas verdienen. Ich und meine Frau und Kinder wollen nicht mehr nach Deutschland, wenigstens nicht um da zu existieren, wohl aber einmal auf Besuch. Ich habe für dieses Jahr 6 und ½ Dollar Staats-, Country- und Kopfsteuer bezahlt, das sind die ganzen Abgaben, denn ich war mit 19 Hundert in der Steuer, jedenfalls die niedrigste Angabe, die ich angeben konnte, weil der Einnehmer bey mir riiert [eruiert] für den Distrikt.

Lieber Bruder, und wenn Du kommst, so mache es Dir so bequem wie möglich, verkaufe alles, was Du kannst. Du schreibst mir aber auch, was Du alles mitbringen sollst, laß Dir Ziehbrotzwieback backen, getrocknetes Obst, Pflaumen und Äpfel, 4–6 Dutzend, Wienerbalsam, aber ächten, 2–3 Pfund Hoffmanstropfen, denn die sind hier sehr theuer. Auch bringe Leinenatlas mit für Hosen und Röcke, Dresdnerzeug für Frauenröcke, den ganzen Satz Centrumbohrer, einige gute breite Hacken, aber nicht mit runden Queren, 3 Stück Harmonikas, ohngefähr 10–15 Gulden das Stück, aus Schönlinde oder wo sonst gute zu haben sind, ein gutes Doppelgewehr kaufe auch. Hast Du noch enge Kleider, so verkaufe sie alle, denn sie sind nicht nach der Mode und auch zu enge. Hier muß alles weit sein, schon die Hosen sind nichts werth, hier kann man keine Träger brauchen; bringe Dir auch ander farbiges Hosenzeug mit für Arbeitshosen; bring mir 2 Dutzend Handtücher mit,

sechsviertel lang; auch ist das Inneltzeug (Inlet) nicht zu haben, bringe auch ein Stück mit, einerlei was für Farbe. Willst Du aber gute Hemden, so lass Dir ein Stück blau und weißen Köper machen, aber nicht feiner als von 30 Wolle, 4 und 4 die Werfte. Auch sind leinerne Bettücher sehr angenehm, denn sie kühlen im Sommer. Da kannst Du auch welche mitbringen, denn unsere sind jetzt zu Ende. Deine Tischtücher bringe aber auch mit, denn hier kannst Du sie brauchen. Kannst Du zinnernes Geschirr kaufen, wie z.B. Teller und Schüsseln, auch Krüge, denn das Porzellan ist hier ziehmlich teuer. Das wäre ohngefähr, was ich Dir rathen kann. Ich verbleibe
Dein Dich liebender Bruder
Christian Friedrich Bergmann

[9.1]
Guadelupe, den 18. Mai [18]67

Lieber Bruder!
Ich habe jetzt einen Auftrag erhalten. Wenn Du nach Amerika kommst, so verwechsel Dein Geld nicht, denn es kommt ein Freund von mir wieder herüber, der will Dir Dein Geld abwechseln. Bey dem bekommst Du es billig und lauter amerikanisches Geld und alles in Gold, da hast Du am wenigsten Verlust. Es ist Hermann Woltersdorf aus Löbau, des alten Richter sein Schwiegersohn, den Du schon kennst. Ich habe aber auch noch etwas mitzubringen, nämlich eine starke zweispanige Winde und eine Schraube zum Hemmen, aber unter den Wagen, denn hinten kann man hier keine brauchen.

Meine Aussicht für eine Ernte ist bis jetzt sehr befriedigend, ich habe zwar nur Mais und ein paar Acker Baumwolle dieses Jahr, ich bin aber auch sehr zufrieden, daß ich keinen Weizen und Roggen gesät habe, denn die Grashüpper waren letzten Herbst in solchen Massen, daß fast gar keine Weizen oder Roggenernte gemacht wird. Und ein Glück, daß wir noch im halben März einen sehr starken Norder bekamen, welcher die jungen Grashüpper wieder tötete. Du sollst auch Friedrich und Karl jedem eine Schraube mitbringen, natürlich mit allem, was dazu gehört zum Anmachen. Woltersdorf wird bis zum 1. Juny hier fort reisen, und so wird es noch Zeit genug sein, daß Du das Geld kannst auswechseln. Woltersdorf ist ein volles Jahr bei mir in der Kriegszeit gewesen, und der alte Richter hat beinahe seine halbe Zeit bei mir verlebt. Woltersdorf wird Dir über vieles Aufschluß geben.
Viele Grüße an unseren Vater und Geschwister, Schwager,
von Deinem Bruder
Christian Friedrich Bergmann

[9.2]
Guadelupe, den 16. Mai 1867 [Brief von Bergmanns Sohn Karl, geboren 1845]

Lieber Großvater und Pathe!
Ihr werdet Euch wohl gewundert haben, daß ich nicht auch an Euch geschrieben habe. Ihr werdet glauben, ich hätte Euch vergessen. Dem ist aber nicht so! Ich war nemlich (wie der Brief abgeschickt wurde) vier Wochen nicht zuhause. Da aber der Vater jetzt wieder schreibt, kann ich die Gelegenheit nicht unbenutzt vorübergehen lassen, Euch ein paar Zellen zu schreiben. Ich stehe nemlich im Begriff, bis zum 26ten dieses Monats zu heirathen, mit einem Mädchen namens Emma Schladör. Ich habe dieses Jahr mit ihrem Pflegevater, einem Deutschen namens Herrter, 25 bis 30 Acker Korn (Mais), 3 Acker mit Weizen und etliche Acker mit Zuckerrohr bepflanzt, und ich erhalte von allem die Hälfte. Wir hatten zehn Acker mit Weizen und 2 [mit] Roggen. Doch die Heuschrecken haben alles bis auf die 3 Acker Weizen abgefressen und hätten auch diese noch gefressen, wenn nicht am 13. März ein sehr kalter Norder gekommen wäre, wo sie alle erfroren. Es war dies ein Glück, sonst wäre gar kein Weizen in der ganzen Gegend gezogen worden. So hat man wenigstens die Aussaat gerettet. Er hat freilich auch viel Schaden angerichtet an den Bäumen, denn es waren schon sehr viele ausgeschlagen, namentlich Maulbeer- und Pfirsichbäume. Es sollte dies auch der kälteste Tag in den Vereinigten Staaten gewesen sein, wo man noch davon gehört hatte.

Lieber Großvater, Euer Bild, welches Ihr mir geschickt habt, habe ich erhalten, und es freut mich sehr, daß Ihr noch so rüstig seid. Auch das Bauerlied habe ich erhalten, welches für jeden Bauersmann eine Ermunterung ist. Ich danke Euch herzlich dafür. Auch wir wollten Euch unser Bild schicken, doch haben wir bis jetzt noch keine Gelegenheit gehabt, uns abnehmen zu lassen, denn der Weg nach San Antonio ist ziemlich weit und darum sehr umständlich. Ich hoffe aber, daß wir es bald können.

Doch ich muß jetzt schließen. Grüßet alle von mir, Onkel, Tanten, Vettern und Muhmen, grüßt auch meinen Pathen C.F. Bergmann auf der Haine. Sagt Dammbitterlich, daß wir zwar Feder und Tinte hätten, aber wenig Geschick sie zu führen, und so hoffe ich auch, daß Ihr mir verzeihen werdet, denn aus dem Grunde hatte ich immer eine Scheu, an Euch zu schreiben. In der Hoffnung, daß Euch diese Zeilen gesund antreffen werden, grüßt Euch
Euer Euch liebender Enkel
Karl Heinrich Bergmann

[10.1]
Guadelupe, den 30. Juli [18]71

Liebe Schwägerin!
Deinen Brief habe ich den ersten Pfingstfeiertag erhalten, aber ich hatte mich sehr gefreut, einen Brief von meinem Bruder zu bekommen, denn er hatte mir in dem letzten Briefe geschrieben, daß ich in kurzer Zeit einen langen, langen Brief sollte erhalten, aber statt diesem war es sein Todtenschein. Wir waren alle sehr betrübt über diese Nachricht. Mir war es, als ob es nicht möglich wäre, aber die Reihe kommt auch an mich. Ich hatte mir vorgenommen, gleich zu schreiben, aber die Arbeit drängte sich massenhaft zusammen, und ich brauche auch eine ziehmlich Zeit, ehe ich einen Brief schreibe. Und so wurde es auch auf den Sonntag aufgeschoben, und jedesmal kamen Leute, welche mich wieder daran verhinderten, bis es endlich auf die lange Bank kam.

Ich kann Dir zwar nicht viel Neues schreiben. Die diesjährige Ernte fällt sehr dürftig aus. Es hat schon seit dem halben Mai keinen Tropfen mehr geregnet, der letzte Regen war ein Wolkenbruch, er ist aber nicht bis zu mir gekommen. Aber wo er war, hat er einen furchtbaren Schaden den Farmern angerichtet. Meine Weizenernte war dieses Jahr eine schlechte, und doch haben andere noch weniger gedroschen als ich. Mit dem Maiskorn ist es bey mir besser wie bei vielen anderen, denn es sind schon in vielen Feldern die Kornstangen für Futter abgeschnitten worden, und das habe ich noch nicht nöthig. Ich werde doch mehr wie eine halbe Ernte machen. Ich habe erst mein Korn nach dem Regen eingesät, und so ist es gekommen, als der Boden nicht so austrocknen konnte.

Liebe Schwägerin, es sollte mich sehr freuen, wenn Du das Guth könntest behalten, denn ich möchte auch nicht gern hören, daß ein Fremder da wohnen sollte. Christoph wird zur Ebersbacher Kirmst Hochzeit haben. Wir haben übrigens sehr viel Arbeit, nähmlich Schindeln schneiden, denn ich schneide auch für die Leuthe, wenn sie Holz her bringen. Ich bekomme für das Tausend einen Dollar, und da mache ich immer noch gut aus. Ich, Friedrich und Christoph können 10 Tausend in einem Tag schneiden. Allerdings halten wir es nicht alle Tage aus, denn es muß auch die Nacht durch gefeuert werden, und übrigens ist es auch zu warm jetzt im Sommer, um tagtäglich bei solchem heißen Holze zu stehen. Nun die Mehrzahl, welche Schindeln schneiden lassen, lassen dieselben auch gleich für den Markt zurecht machen, und da bekomme ich einen Dollar und einen halben für das Tausend, und so wird dann mit der Arbeit gewechselt, einmal Schneiden und einmal Packen, und so halten wir es besser aus. Ich kann Dir sonst keine Neuigkeiten schreiben. Zwar sind

in letzter Zeit die Indianer sehr schlimm gewesen und haben viel Pferde gestohlen, aber ich habe bis jetzt noch nichts von ihnen zu leiden gehabt. Es sind aber auch seit kurzer Zeit Schritte von der Regierung gethan, welche auf einen Indianerkrieg mit Bestimmtheit schliessen lassen, und so sind auch die Grenzansiedler wieder guten Muthes, denn sie haben 3 Jahre lang viel Pferde gestohlen und auch Menschen getöthet.
Grüß mir Deine Kinder von Ihrem Onkel und Deinem Schwager
C.F. Bergmann

[10.2]
[Brief von Bergmanns Frau Luise]

Liebe Schwägerin!
Ich kann nicht unterlassen, Dir ein paar Zeilen zu schreiben, da ich doch auch gerne etwas von meinen Verwandten und Freunden wissen möchte, weil sie doch auch Deine Verwandten sind. Wenn Du uns wieder schreibst, so schreibe mir, was Deine Eltern und Geschwister und Deines Vaters Geschwister machen. Denn von der Gabriel Freuden habe ich noch nichts gehört, die hat mich wohl ganz vergessen; ich habe doch einigemal ihrer erwähnt in Briefen, auch die andern.

Die Nachricht von meines Bruders Tode kam mir sehr erwünscht, denn auf Besserung durfte ich nicht hoffen. Er war zu weit gesunken, da er nun der Gemeinde zur Last fiel, wo mein Vater schon großen Kummer darum hatte. Denn der sah es schon voraus, daß es so kommen mußte. Grüße alle von mir, auch die Heinrich, Randig und Frau, Dammbitterlichs K. und Familie und Schwester Eleonore und Herrn Organist Michael. Ich bin noch gesund und wünsche, daß Dich auch diese Zeilen gesund antreffen möchten. Schreib uns sobald als möglich wieder.
Ich verbleibe Deine Schwägerin in der weiten Welt, J. Ch. Luise Bergmann

[11.1]
[kein Datum; wahrscheinlich zwischen 1870 and 1874]

Lieber Vater, Brüder, Schwestern u.s.w.!
Lang ist es her, daß ich, wie Ihr sagt, nicht geschrieben habe. Dem ist aber nicht so. Ich habe Euch den vorletzten Brief beantwortet und gesagt, daß Ihr dem Schmit die Zinsen sollt geben und daß ich sobald als möglich das Geld schicken werde. Auch wollte ich gleich den Brief aus Kalifornien beantworten, aber ich brachte das Geld nicht zusammen, und so hat es sich

einen Monat länger verzogen, als ich wollte. Es hat mir schon viel Sorgen gemacht, daß ich es nicht konnte, aber allerhand Verluste zwangen mich, daß ich es nicht konnte. Ich habe für die Wagen Fracht, Spesen und Zoll 123 Doll. in Indianola bezahlen müssen, und dann habe ich auch hier noch ziehmlich Schmiedekosten gehabt. Das wäre aber alles gut gewesen, wenn nicht mit der ersten Ladung der Wagen von dem neuen Stellmacher zerbrochen wäre. Wir waren erst 3 Meilen von Hause, da brachen zwei Räder, ein vorderes und ein hinteres. Das war ein Spaß von 46 Doll., denn ich konnte bloß die hinteren Ringe gebrauchen, die anderen waren zu klein. Die Böcke waren zu schwach. Letztes Frühjahr ging auch der andere entzwei, den hatte ich dem Karl gegeben. Karl ist auch verheirathet, er wohnt aber 30 Meilen weiter oben in Comfort.

Nun will ich Euch aber noch sagen, daß ich seit 4 Jahren 27 Zugochsen verloren habe, theils durch Krankheit, theils gestohlen und fortgelaufen, theils auch so verunglückt. Darunter waren 7 Stück, wo ich für jeden 25 Doll. bezahlt habe und das hat mich freilich etwas zurück gebracht. Im letzten Herbst war es jährlich, wie Ihr wisst, daß Reimann zu mir kam. Er konnte nicht gleich Arbeit bekommen, und so habe ich ihn bei mir behalten, bis er Arbeit hatte. Er ging dann zu einem Nachbarn, wo er arbeiten konnte. Daselbst war er 3½ Monat, dann entließ ihn derselbe, weil er zu grob war. Dann kam er wieder zu mir und überredete mich, eine Schindelmaschine zu bauen, und ich ging den Vorschlag ein (nämlich in Compagnie), und so haben wir 3½ Monat gebaut, bis die Maschinerie fertig war. Ich war der Geldaufleger. Während dieser Zeit hat er bei mir Kost und Wäsche gehabt. Endlich war die Maschine fertig, und wir fingen an zu schneiden. Wir hatten 16 Tausend Schindeln geschnitten und wollten sie in Pakete machen, aber da wurde es anders. Es fing an zu regnen und regnete 4 Tage ununterbrochen, und unsere Schindeln gingen fort. Das war wieder ein Schaden von 110 Dollar. Dann war aber auch die Maschine mit fort und Reimann dazu. Und nun, Bergmann, sieh zu, von wem Du für Deine Auslagen etwas bekommst.

Bei diesem Regen war der Guadelupe 53 Fuß! gestiegen. Ich hatte das Wasser in der Küche. Ich hatte ein gemauertes Rauchhaus, das hat sich bis auf die letzten 3 Stunden gehalten, aber da kam eine große Zypresse und machte es dem Erdboden gleich. Auch habe ich viel Schaden im Felde gehabt. Es hat mir über zweitausend Riegel mit fortgenommen, von 6 Ackern den ganzen guten Boden mit fort gewaschen, nebst dem Korn, welches darauf stand. Aber auch das andere hatte viel gelitten, indem das Wasser mehrere Tage hoch blieb, so daß ich nur gut die halbe Ernte gemacht habe. Aber ich tröste mich damit, daß ich es nicht selbst verschuldet habe und daß es anderen auch so gegangen hat. Alle, welche an größeren oder kleineren Flüssen gewohnt haben, haben auch alle mehr oder weniger Schaden. Es hat die ganzen Mühlen mit fortgenommen, kurz, alle Gewerke sind theils ganz ruiniert, theils viel beschädigt. Ich habe aber vergessen zu sagen, wieviel ich für die Schindelmaschine an bar ausgelegt habe. Es waren 94 Doll., die ich erst für den Schmit bestimmt hatte, ehe ich zu dem Entschluß kam, die Maschine zu bauen. Ich wollte über Reimann kein Wort sagen, wenn er zu mir gesagt hätte, ich will keinen Antheil mehr an der Maschine haben. Aber er ging von mir fort nach Boerne in die Mühle arbeiten, wo er mir 5 Doll. gegeben hat, von dort nach New Braunfels und ist nicht wieder zu mir gekommen, und so habe ich auch da wieder das Nachsehen. Er hat sich, während er in Boerne und New Braunfels gearbeitet hat, schönes Geld verdient, aber ich weiß nicht, wo es bleibt. Er hatte noch einen Sattel bey mir. Den hat er unter der Hand verkauft, so daß ich auch den nicht haben sollte.

Ich muß noch einmal auf das hohe Wasser zurückkommen. Es hat mir sehr viel Schaden an den Bäumen gemacht, welche längs der Guadelupe standen. Viele Schindelbäume hat es zerbrochen und mit fortgenommen, Bäume von 3½ Fuß Durchmesser, andere mitten zerbrochen. Ich hatte eine Pappel, welche nahezu 4 Fuß Durchmesser hatte. Auch diese ist fort, denn was das Wasser nicht fertig brachte, das brachte das Treibholz fertig, und so kann ich Euch den Schaden nicht angeben.

Ich und Friedrich waren kurz vor dem hohen Wasser in meinem Zederwalde und beschlugen Holz für sein Haus. Es ist nicht Mode, daß man das Handwerkzeug mit nach Hause nimmt, und wie wir wieder über die Guadelupe konnten und wieder in den Wald konnten, war aber auch alles Handwerkszeug weg. Denn es vergingen neun Wochen, ehe wir wieder über die Guadelupe konnten. Zum Glück waren die ganzen Kühe an dieser Seite. Sonst hätten wir nicht einmal Milch und Butter gehabt. Ganz anders war es mit den Zugochsen. Da war auch nicht einer auf der Seite. Das machte aber das Maß der Arbeit voll, die Fens war offen, und kein Zugochse darin, und nach dem Regen drückende Hitze und mehr Arbeit, als mir lieb war. Dazu noch kränklich und keinen Arbeiter zu bekommen, denn es hatte ein jeder selber genug zu thun. Und so haben wir drei, ich, Friedrich und Christoph, Reisig herbei geholt, um nur die Fens zu machen. Aber demohngeachtet kam doch noch Vieh ins Feld, denn die Guadelupe fiel in den ersten Tagen so schnell, so daß immer wieder des Nachts das Vieh ins Feld kam. Ich habe da viel Schaden von dem Vieh gehabt. Ich habe so nur eine halbe Ernte. Dieses Jahr habe ich wieder Roggen und Weizen gesät, 6 Acker.

Über unsere politischen Verhältnisse denken wir bald besser sprechen zu können. Wir haben die Demokraten in dem Wahlkampfe geschlagen, der neue Gouverneur [ist] ein guter Unionsmann; auch die Kongressmitglieder sind radikal, so wie auch die in dem Senat. Es sind alle, die ich kenne, tüchtige Männer, und so hoffen wir, daß auch der Staat Texas wieder in die Union aufgenommen wird.

Es wird nächste Pfingsten zwei Jahre, da schickte ich Christoph nach Indianola. Ich gab ihm 28 Doll. Geld mit, er war 8 Wochen unterwegs, kam krank nach Hause, brachte 53 Doll. Geld mit und hatte ein Joch Ochsen verloren. Er war 4 Monate krank, natürlich wurde auch die Zeit nichts verdient. Ich selber kann nicht mehr fahren, denn es ist natürlich nicht wie in Deutschland, daß man die Zugthiere im Stalle hat des Nachts, und ich kann wohl noch fahren, wenn ich einen bey mir habe, der mir die Ochsen morgens hilft suchen, denn seit ich nur noch auf ein Ohr höre, ist das Suchen vorbei. Ich kann mich auf eine Seite wenden, wie ich will, so höre ich auch die Glocken von der nähmlichen Seite, wo ich mich hin wende. Ein Beispiel will ich nur anführen: Ich will also die Pferde suchen. Ich höre die Glocke immer, komme aber immer weiter davon ab, so daß ich mich genöthigt sehe, wieder nach Hause zu gehen, nachdem ich stundenlang war herumgelaufen. Sodann bin ich wieder zu Hause und habe mir meine Frau mit genommen (natürlich mochten die Pferde wohl nicht auf derselben Stelle geblieben sein), und so könnte ich noch mehr solche Abentheuer erzählen.

Lieber Bruder, Du hast mir schon geschrieben, wegen einem Golddollar. Ich habe zwei eingelegt, einen für Dich und einen für meinen versprochenen Pathen. Ich habe schon 5 solche Dinger eingelegt, aber leider sind die Briefe auch nicht angekommen. Dieses Mal will ich aber den Brief registrieren lassen, um sicherer zu fahren. Ich hätte auch gerne meine und der Frau ihre Photographie mitgeschickt, aber es hat noch nicht gepaßt, daß wir alle beide nach San Antonio konnten. Einmal waren wir dort, aber da war der Photograph nicht da und so ist es immer geblieben.

Noch eins, ich habe mich schon mehrere Jahre mit einem Bandwurm geschleppt und kann ihn nicht loswerden. Ich habe drei Kuren gemacht, aber allemal erfolglos. Sobald ich aber werde Zeit haben, will ich nach San Antonio und ihn mir lassen forttreiben. Ich habe immer die 10 Doll. gescheut, und es kostet mich auch eine Woche Zeit, aber ich muß jetzt daran gehen. Sonst sind wir alle noch gesund. Die drey Jungens sind etwas verkümmert. Friedrich ist beinahe so groß wie ich, Karl ist zwei Zoll länger wie ich, und Christoph ist einen halben Zoll kleiner wie Karl. Ich glaube aber, daß er ebenso groß wird

wie Karl. Ich will die Littaney schließen, ich bin jetzt, denke ich, über den Berg.
Euer [Euch] liebender Sohn und Bruder und Schwager
Christian Friedrich Bergmann

Vollständige Adresse:

Abs.: C. F. Bergmann
P.O. Spring Braensch
Comal Co.
Texas

Wenn Ihr einen Brief schickt, so adressiert ihn nach Comal Co., P.O. Spring Braensch, es ist eine neue Postoffice und ich habe nicht weit dorthin.

[11.2]
[Brief von Bergmanns Frau Luise]
An Christiane Elisabeth Bergmann, Witwe
Den 6. Dezember [kein Jahr, wahrscheinlich zwischen 1870 und 1874]

Liebe Schwägerin!
Deinen Brief mit den Familienereignissen haben wir erhalten und haben Dir auch wieder geantwortet. Auch hatte Christoph Dir seinen Dank für das Geschenk darin ausgesprochen; wir schickten damals zwei Briefe fort, einen an Dich, den anderen an August gerichtet.

Nun versprachst Du, wenn Karl zu den Ferien nach Hause käme, so sollte er uns auch wieder schreiben. Nun haben wir immer gewartet, aber vergebens. Sollte es sein, daβ er bald nach Hause zum Besuch käme, so seid so gut, und schreibet uns auch ein paar Zeilen, daβ wir auch wieder was hören. Ich kann Dir weiter nichts schreiben, das andere hat mein Mann alles geschrieben.
Viele Grüβe an alle Verwandte und Bekannte
Joh. Chr. Luise Bergmann

[12.1]
Guadelupe, 4. Januar 1879

Liebe Schwägerin!
Ich habe lange auf den Brief warten lassen. Ich habe immer auf die Photographie gewartet, es ist aber bis heute noch keine gekommen. Und noch

mehr, ich bin aber auch kein Briefschreiber, denn es will immer lange Zeit, ehe ich die Feder in die Hand nehme, um einen Brief zu schreiben. Oft fehlt es an Zeit, oft an Lust. Ich habe 20 Acker Land zu bestellen und bin nur allein. Da muß ich dann doch immer etwas auf dem Trab sein. Das beste ist, daß ich und meine Frau so ziehmlich gesund sind.

Liebe Schwägerin, Du schriebst mir die Preise der Nahrungsmittel bei Euch, und so will ich auch Dir die Preise schreiben, was sie bey uns kosten. Ein Bushel Weizen von 60 Pfund kostet 60–70 Cent. Winter Weizen 1 Doll., Roggen 50 Cent, Hafer 25–30 Cent, Mais 30–35 Cent. Rindfleisch 5–6 Cent. Die Butter im Sommer 10–15 Cent, jetzt 30–40. Eier im Sommer 10–15 Cent, jetzt 20–30. Schwein geschlachtet, auf dem Market 3–9½ Cent das Pfund. Ein 3 Jahre alter Ochse kostet 10–11 Doll., 4 Jahre 12–14 Doll. Kartoffeln das Bushel 1 Doll. und 30 Cent., Bier 5 Gallonen 2 Doll. 25 Cent. Branntwein die Gallone 1 Doll. 50–2 Doll.; Kaffee 5 Pfund, Zucker 9–10 Pfund, Hutzucker 8 Pfund für 1 Doll. Ein Knecht erhält 12–15 Dollar monatlich. Ein Mäher bekommt für den Acker 1 Doll., so auch die Maschine. Die Kleiderstoffe sind auch sehr billig. 1 und ½ Elle, das heißt ein Yard, kostet von 16–24 Yard 1 Dollar. Hosenzeug 15–25 Cent, Schuhe 1½ Doll., für Frauen 75 Cent bis 1 Dollar, auch 1½. Stiefeln 3–4 Doll.

Ich habe dieses Jahr keine gute Ernte gemacht, ich hatte 5 Acker mit Weizen gesät und habe 18½ Bushel gedroschen, Witterungsverhältnisse waren die Ursache. Der Hafer war gut, aber der Mais war auch nur eine gute halbe Ernte. Baumwolle war eben auch nicht zum besten. Die trockene Witterung hatte ihr viel geschadet, und nun hat sie auch einen sehr niedrigen Preis. Sie ist in San Antonio vor Weihnachten für 5–6 Cent das Pfund verkauft worden. Ich will aber meine jetzt nicht verkaufen, sie fängt schon wieder an zu steigen.

Von dem Gelben Fieber, welches so furchtbar in New Orleans und in vielen anderen zum Theil großen Städten gewüstet hat, die Sterblichkeit war so groß, wie sie noch nie bei Gelbfieberfällen war, es ist haarsträubend, wenn man die Berichte in den Zeitungen liest. Auch der Export und der Import haben bei uns viel durch die Quarantäne zu leiden gehabt.

Unsere Verhältnisse in Texas sind sehr betrübender Art. Man kann irgendeine Zeitung in die Hand nehmen, wie man will, so sind sie voll von Raub und Mord. Die Pferde oder Ochsen werden oft in der Nacht gestohlen. Auch kommt es sehr häufig vor, dass einem der Erlös von seinem Erwerb unterwegs gestohlen wird, wenn nicht gar totgemacht. Das ist aber nicht bei uns und im Gebirge der Fall. Die Post und die Eisenbahn werden angehalten und beraubt, und das ist in letzter Zeit öfter der Fall gewesen. Das Pferde- und

Ochsenstehlen betreiben in der Regel die Mexikaner. Diejenigen, welche die Post und Eisenbahn berauben, sind aufwendig große Herren.

Wir haben dieses Jahr einen kalten Winter, es ist zwar nicht so grimmig kalt, aber es ist fortwährend Norder (Nordwind) und dabei trübe und auch naß, so dabei der Boden in den Feldern so trocken ist, daβ viel, sehr viel Weizen nicht gesät werden konnte, und jetzt, wo die Zeit für Sommerweizen kommt, ist der ewige Norder. Dabei leidet aber auch das Vieh. Ich und Friedrich füttern dasjenige, welches zu Hause kommt, mit Korn und Baumwollsamen und Haferstroh. Letztes Frühjahr habe ich ein Kalb gekauft für 15 Dollar (ein Ochsenkalb).

Liebe Schwägerin, Du beklagst Dich, daß ich lange nicht geschrieben habe, ich habe jeden Brief beantwortet. Samuel schrieb mir und bat mich um Samen, welchen ich ihm auch geschickt habe, und ich bat ihn, er sollte mir das Resultat schreiben, aber ich habe keine Antwort erhalten. So auch mit Gustav Dreßler, dem habe ich auch geantwortet. Sei es nun, daß die Briefe verloren gegangen sind, oder Du hast nichts davon erfahren. Friedrich hat eine Farm gekauft für 500 Dollar. Diese grenzt an unsere Farmen an. Er hat jetzt 4 Kinder, 3 Buben und 1 Mädchen. Sie sind alle gesund. Christoph hat auch 4 Kinder, 3 Buben und 1 Mädchen. Karl hat keine Kinder. Christoph ist auch mit seiner Familie gesund, aber Karls seine Frau ist immer etwas krank, er ist gesund.

Ich habe oben an einer Stelle die Zustände in Texas geschildert, die freilich nicht sehr einladend sind, aber ich möchte nicht wieder in Deutschland sein. Wir werden von alledem nicht beeinträchtigt, was ich oben gesagt habe. Der Hochmuth und Nahrungssorgen drücken uns nicht, trotz der schlechten Ernte, die wir hatten. Ich habe letztes Frühjahr für 180 Doll. Vieh verkauft und kann auch dieses Frühjahr wieder ziehmlich verkaufen. Ich habe dieses Jahr 4 große Schweine geschlachtet und auch eine Kuh, wovon ich sehr viel Bratwurst und auch andere Wurst gemacht habe. Ich habe schon manchmal wieder meiner Frau gesagt, wenn nur die Leuthe in Deutschland das zu essen hätten, was wir nicht wollen. So wird zum Beispiel der Kopf und die Füße von einem Ochsen nur für den Hund hingeworfen, das andere Eingeweide für die Schweine. Sehr oft wird nicht einmal das Talg von den Därmen abgemacht. Wenn Du unsere Kühe einmal solltest inspizieren, so würdest Du wohl sagen, das sind doch sehr liederliche Leuthe. Wir haben jeden Sonntag unseren Kuchen, nicht bloß wir; man kann hinkommen wo man will, es ist Kuchen da. Gekochtes Fleisch schmeckt nicht, es muß gebraten sein, bloß wenn Suppe sein soll, wird welches gekocht.

Ich verbleibe Dein Schwager
C.F. Bergmann und
L. Bergmann

Viele Grüße an meine Schwestern und Schwäger!

[12.2]

Liebe Schwägerin!
Ich will Euch auch ein paar Zeilen schreiben, daß Ihr glauben könntet, ich bin noch auf der Welt und wünsche Euch allen viel Glück zum Neuen Jahr aus der Ferne! Mein Mann hat Euch die Schattenseiten und auch das andere getreu geschildert, und so jemand Lust hat zu kommen, so soll er unverzagt kommen. Texas ist noch groß, und ordentlichen Leuten wird überall auch fortgeholfen.

Die erste Zeit war's Heimweh groß. Jetzt möchte ich mit Euch nicht mehr tauschen, in keiner Beziehung, und so geht es allen die hierher kommen. Auch die zurück nach Deutschland gegangen sind, kommen bald zurück. Es kommen jetzt auch viele Emigranten aus den nördlichen Staaten nach Texas. Denn es ist noch viel Land, wo angesiedelt kann werden. Friedrich hat die Farm für 500 Dollar gekauft, 320 Acker. Die Gebäude darauf waren zwar schlecht und die Fenz, doch war sie billig. Um unsere Gegend giebt es mehr Deutsche diesseits der Guadelupe, und Amerikaner verkaufen gern. Sie lieben nicht so sehr, unter den Deutschen zu wohnen und umgekehrt mit den Deutschen. Wir haben auch Eisenbahnen in Texas, von Galveston bis San Antonio. In einem Tag fahren sie bis San Anton von Galveston. Auch ist eine von Houston über Austin gebaut.

Ich will schließen mit dem Wunsche, daß Euch diese Zeilen alle gesund antreffen mögen. Viele Grüße von mir an alle Verwandten, Bekannten und Freunde, und sage, es ginge mir gut. Lebet wohl, ich verbleibe Eure Schwägerin
Joh. Chr. Luise Bergmann

[12.3]

Lieber Neffe!
Ich habe aus Deinem Schreiben gesehen, daß Du auch Schulmeister werden willst, von mir aus hast Du Recht und [ich] wünsche Dir auch viel Glück. Aber was Du von Amerika kommen schreibst, das ist mir beim Lesen wie ein Traum vorgekommen. Nun ja, so komm, von Galveston bis San Antonio ist nur eine Tagesreise, und von San Antonio bis zu uns ist auch nicht mehr. Aber mache

Dich etwas mit der englischen Sprache bekannt, sonst wirst Du, wie jeder andere auch, als Grünhorn angesehen. Dein Vater wollte mich schon hier in Amerika besuchen, aber der Tod kam zu früh. Was Du mitbringen sollst, so wirst Du mir wohl erst einmal schreiben, überhaupt, wenn es noch so lange Zeit hat. Du mußt aber auch bedenken, daß ich nächsten April den zweiundsechzigjährigen Geburtstag feiere, und meine Frau hat den einundsechzigjährigen vor Weihnachten gefeiert. Es wird mir die Zeit doch lang werden, bis ich Dich sehe. Aber ich hoffe doch, Dich noch zu sehen.
Dein Onkel
C.F. Bergmann
Guadelupe, Kendall County

[12.4]

Lieber Neffe und Pathe!
Wie Du mir geschrieben hast, haben sich Deine Verhältnisse wesentlich geändert, und ich freue mich auch, daß es Dir gut geht. Du wirst auch gerne viel aus Texas wissen wollen, aber ich habe mich schon in dem Briefe an Deine Mutter erschöpft.

Ich hätte es gerne gesehen, wenn ich Dich in Texas gesehen hätte, aber jetzt muß ich den Gedanken doch aufgeben. Ein Lehrer in San Antonio hat von 57–100 Dollar im Monat und auf dem Lande von 20–30 und 40, je nachdem er die Kinder hat. So viel Kinder darf ein Lehrer nicht haben wie in Deutschland. Sie müssen aber auch den ganzen Tag in der Schule sein. Dieselben, welche weit von der Schule wohnen, müssen ihr Mittagbrot mit in die Schule nehmen.

Es sind auch hier in jedem größeren Sättelment Gesangvereine und, wie sich Sachverständige aussprechen, auch was Gutes leisten. Sie haben alle Jahre ein großes Gesangfest. Das letzte war in San Antonio, welches 3 Tage dauerte, und die Gesangvereine aus Galveston, Houston, Austin, Friedrichsburg, Comfort, Boerne u.a.m. mit Theil nahmen. Auch waren zwei Musikchöre, welche abwechselnd spielten. Ich kann Dir weiter nichts Neues schreiben, Du wirst doch jedenfalls den Brief von Deiner Mutter lesen, in welchem ich mich, so gut ich konnte, ausgedrückt habe.
Viel Glück zum neuen Jahr
Dein Onkel und Pathe
C.F. Bergmann
Guadalupe, Kendall County

[13.]
Guadelupe, den 28. May 1880

Lieber Neffe!
Ich habe Dich lange auf einen Brief warten lassen. Ich wollte Dir recht viel schreiben, aber es hat sich nicht viel Besonderes zugetragen. Mir ist aber auch der Pflug und die Hacke händiger als die Feder. Ich brauche lange Zeit, ehe ich einen Brief geschrieben habe, und dann noch mit zitternder Hand, wie Du aus diesem Briefe sehen kannst. Denn ich bin zum nächsten April 63 Jahre alt und das Schreiben kommt mir nicht oft vor. Deinen Brief habe ich erhalten mit den Photographien und bedanke mich auch für dieselben. Es hat mich sehr gefreut, das Haus meiner Eltern wiederzusehen, aber auch über den Ostritzer Organist habe ich mich sehr gefreut auf der Photographie. Der Todt Deiner Mutter kam uns aber auch sehr unerwartet. Was meine Schwester angeht, so wollte ich, daß sie den Hintern ganz gehörig gewaschen bekäme.

Wir haben hier in West-Texas dieses Jahr soviel wie keine Ernte gemacht, und es hat von Weihnachten bis September nur zweimal geregnet, so daß kein Weizen gesät konnte werden, und diejenigen, welche gesät hatten, ging der Samen nicht auf. Ich habe zwar soviel geerntet, wie ich für meine Wirthschaft brauche, aber zum Verkaufe habe ich nichts übrig. Du brauchst Dir aber nicht einzubilden, daß wir Noth leiden, ich habe vier Schweine und eine Kuh geschlachtet und wollte nur, daß die armen Leute das hätten, was bei uns die Hunde zum Fressen bekommen. Denn der Kopf, Herz und Lunge, sowie die Füße und die Flecke werden von dem Rindvieh nicht gebraucht. Kurz und gut, wir hungern nicht. Ich kann auch allen Tag mein Glas Wein trinken, ich habe dieses Jahr 200 Quart [1 Quart = 1,136 liter] gemacht. Freilich, es ist kein Rheinwein oder Burgunder, aber er läßt sich doch trinken.

Von Deutschland haben wir schon vergangenen Sommer betrübende Nachrichten erhalten, von dem vielen Regen, welcher überall gewesen ist. Nun habe ich in den Zeitungen vom 16. Januar gelesen, daß in ganz Europa der Eisgang so ungeheuren Schaden angerichtet hat. Auch in Amerika ist viel Schaden durch die Elemente angerichtet worden, in vielen Staaten sind Tornados gewesen, welche ungeheuren Schaden angerichtet haben. Wir haben schon im vergangenen Herbst von dem vielen Schnee und Kälte gelesen, welche in Deutschland war. Und nun die politischen Wirren und der Hunger in Schlesien und theilweise in Sachsen.

Auch wir hatten zu Weihnachten große Kälte, das Thermometer war auf vielen Plätzen bis auf 17 Grad unter Gefrierpunkt gefallen, Es sind auch eine ganze Menge Menschen erfroren, welche sich theils zu Weihnachten einen [?]

gekauft und auch welche auf der Reise waren. Ich hatte 4 Acker Winterhafer gesät, aber der Frost hat ihn total vernichtet. Übrigens haben die Farmer wieder viel gesät. Sie konnten aber erst nach Weihnachten daran denken, es war immer zu trocken zum Pflügen.

Halt, was ist das? Eine Postkarte. Du hast Recht, daß Du Dir eine Antwort auf diesem Wege forderst. Ich war gerade mit dem Briefe so weit, als ich dieselbe erhielt. Unsere Ernteaussichten sind gut, so daß wir für das letzte Jahr wieder entschädigt werden, wir fangen an, Getreide zu mähen. Den 11. April ist Karl seine Frau gestorben. Sie hatte 7 Monate gelegen.

In Amerika sind Waldbrände und Feuersbrünste die stehenden Neuigkeiten. Es sind im Staate 50.000 Acker Waldung verbrannt, so wie die nahe gelegenen Dörfer, wobei auch eine ganze Anzahl Menschen. Du darfst es mir nicht übel nehmen, daß ich keinen Brief an Dich geschrieben habe, und wo sollte ich auch den Stoff hernehmen. Lügen will ich nicht. In 14 Tagen wird die große Aloe bei mir blühen, da wünschte ich doch, daß Du sie sehen könntest. Die Blumenstange ist jetzt 14 Fuß hoch und bringt erst die Blütenknospen und hat einen Durchmesser von 7 Zoll.

Du wolltest auch wissen, ob ich mit meinen Kindern oft zusammen komme. Den Friedrich sehe ich alle Tage, denn es sind nur 600 Schritte von mir bis zu ihm. Auch mit dem Christoph kommen wir öfters zusammen. Der wohnt nur 10 Meilen von mir. Freilich, mit dem Karl ist es anders, der 33 Meilen bis Comfort entfernt ist. Er war diese Pfingsten zum ersten Male seit vorletztem Weihnachten hier. Jetzt, da seine Frau todt ist, hat er mir versprochen, öfters zu kommen. Und so war es immer das Gegentheil, daß wir zu ihm mußten.

Du willst wissen, wie lange wir bis San Antonio brauchen. Da habe ich einen Tag hin und einen zurück. Die Briefe und Zeitungen müssen wir selber holen. Wir sind jetzt seit einiger Zeit bequem, denn wir haben jetzt eine Post Office in unserer nächsten Nähe. Also, wenn Du wieder schreibst, so ist die Adresse P. O. Guadelupe, Kendall Co., Texas.

Ich habe mir vorgenommen, wenn ich wieder einen Briefe schreiben soll, daß ich von jetzt an alles zuvor zu Papiere bringe, wenn ich an etwas denke, was ich schreiben will. Meine Gedanken werden kurz. Ich hoffe, daß keiner von Euch beiden es wird übel nehmen, daß ich nicht jedem einen Brief geschrieben habe, da ich doch in jedem ein und denselben Inhalt hätte schreiben können, und auch die Überzeugung habe, daß keiner von Euch den Brief dem anderem wird vorenthalten.

Wir hatten am 28. May wieder Hochwasser. Die Guadelupe war wieder in 6 Stunden 16 Fuß gestiegen, in der Nacht vom 27sten 12 Uhr bis zum 28sten

Mittag. Sie hat aber keinen Schaden angerichtet. Freilich, der starke Regen hat sehr im Feld die Ernte in die Guadelupe geführt.
Ich will schließen und wünsche, daß Ihr den Brief gesund erhalten möget.
Es grüßt Euch beide
Euer Onkel
Ch. F. Bergmann
L. Bergmann

[14.]
Guadelupe, 8. Dezember 1881
Kendall County

Liebe Neffen!
Ich hatte mir schon eine Postkarte gekauft, um Euch um Antwort zu bitten, als ich Eure Karte erhielt. Ich hatte doch lange einen Brief an Euch geschrieben und keine Antwort darauf erhalten. Nun, ich kann Euch auch heute nicht viel schreiben. Wir beide sind, Gott sei Dank, recht gesund, und ich habe auch eine so leidliche Ernte gemacht. Es sind freilich nicht viele, welche mit Korn so gut ausgemacht haben wie ich. Viele haben die Arbeit gespart, und andere konnten wegen Trockenheit nicht mehr im Korn arbeiten. Es war sozusagen unerträglich heiß und trocken.

Karl hat letzten Sonntag wieder geheirathet, ein tüchtiges Farmermädchen aus Boerne. Da ist denn auch tüchtig gesungen und geschlungen worden. Karl ist auch ein Mitglied beim Comforter Gesangsverein. Ich war auch bei dem letzen Gesangfeste in Boerne. Es waren 7 Gesangsvereine beisammen und dauerte das Fest 3 Tage und 3 Nächte. Was da alles ausgeübt worden, das könnt Ihr Euch denken. Musik und Gesang wechselten immer ab. Den letzten Abend war ein Fackelzug und dann wieder Commers. Das schönste bei solchen Festen ist, daβ kein [?] und kein Armer zu sehen ist, alle in gleichen Schuhen.

Lieber Neffe, Du möchtest gerne wissen, ob auch so getanzt wird wie in Deutschland. Jawohl, all die Tänze wie bey Euch und noch andere mehr, nur muß jeder, der tanzt, Eintritt bezahlen, und zwar verhältnismäßig je nach der Zahl der Musikanten, 50 Cent bis 1 Dollar. Die Musik beginnt aber schon 3 Uhr nachmittags und dauert bis zum nächsten Morgen. Wir waren auch bei dem letzen Mayfeste, ich und meine Frau, und haben uns auch recht lustig gemacht, denn es ist dann doch eine Art Bedürfnis, daß sich alte Leuthe auch einmal austoben, und ich bin der Letzte ganz gewiß nicht, der es versäumt. Ich kann mir auch eher ein Vergnügen erlauben als vor 12–15 Jahren, denn

Interessen [Zins] nehmen oder Interessen geben, ist doch ein großer Unterschied. Friedrich, der älteste Sohn, ist mit seinen Verhältnissen auch sehr zufrieden. Er hat fast alle Ackergeräthschaften, welche ein Farmer braucht, eine Erntemaschine hat er letzen Sommer gekauft, und eine Dreschmaschine wird er zum nächsten Sommer kaufen. Dann hat er alles, was eine gut eingerichtete Farm braucht. Auch ist es ihm gelungen, einen Brunnen zu graben. Er kostet wohl viel Geld, denn er mußte 38 Fuß alles ausschießen. Jetzt ist er aber doch froh, daß er Wasser hat. Auch Christoph ist in ziehmlich guten Verhältnissen. Freilich ist er noch ein jüngerer Farmer, aber er ist auch fleißig, und so wird er auch alle Jahre besser ab. Er verdient sich auch manchen Dollar in seiner Schmiede. Er hat sich selbst einen Blasebalg gemacht und eine alte Schmiede aufgekauft. Beim Karl hoffen wir alle, daß eine neue Aera eintreten wird, nachdem er geheiratet hat. Er ist mit seiner ersten Frau freilich nicht sehr vorwärts gekommen, denn die Doktor- und Apothekerkosten nahmen immer den größten Theil von den Einnahmen, die er machte; und so hat es uns alle gefreut, daß er sich an ein tüchtiges Farmermädchen verheirathet hat.

Mit dem Weinmachen war es diesen Sommer nichts. Er erfror bei dem letzten Froste im Früjahr, und so müssen wir uns nur welchen denken. Jetzt kommt wieder die Zeit, wo wir die Schweine wieder schlachten, und ich habe dieses Jahr nur 3 Stück zum Schlachten, aber sie sind ziehmlich groß, daß ich zwei davon verkaufen kann. Aber es soll wieder tüchtig Wurst gemacht werden. Das Rindvieh hat jetzt einen guten Preis, ein 1-Jahr altes Kalb kostet 40 Mark, ein 2 Jahr altes 60 Mark, große Schlachtochsen sind sehr gesucht und nicht zu finden, eine fette Kuh 72 Mark, ein Schlachtochse 100 Mark, ein Kuh mit Kalb von 72–100 Mark. Ich will bis zum nächsten Frühjar 10 Kühe mit Kälbern verkaufen. Das giebt wieder einmal Biergeld, denn Bier trinke ich gerne.

Ein Abenteuer muß ich Euch doch noch schreiben. Nach der Hochzeit bin ich mit meiner Frau mit nach Comfort gefahren, natürlich in der besten Stimmung. Wir nahmen das nöthige Naß mit, denn Wirtshäuser giebt es bis dahin keine. Endlich zur Stelle angelangt, wo wir unser Naß befreien wollten, wurde angehalten, und es wurde auch befreit. Nachdem wir unsere Schuldigkeit gethan hatten, fuhren wir fort, aber das Schicksal, nähmlich das Hinterrad, wollte eher in Comfort sein als wir, und es ging gerade Berg runter, und so lief es vordern mit einem Stück Schenkel nach, und wir sollten mit drei Rädern fahren. Da gab es wieder ein lautes Hurra für Texas. Jetzt keine Axt beim Wagen, aber es dauerte nicht lange, so war auch eine Axt zur Hand, und in einer Stunde fuhren wir wieder los. Eine Stunde später brach der Langbaum,

wieder Aufenthalt und Lachen. In Comfort angekommen wurde ich mit lauter Lachen bedauert. Wir hatten nämlich eine Schleife gemacht und so halb Schlitten gefahren. Jetzt sind wir wieder zuhause und gehen unserer Arbeit nach. Ich habe auch 25 Dollar Prämie auf der letzten Ausstellung erhalten. Ihr werdet wohl meinen Namen in der Farmer-Presse gefunden haben. Das übrige findet Ihr in der Presse.
Es grüßt Euch Euer Onkel C.F. Bergmann und L. Bergmann
und alle 3 Söhne

[15.]
Guadelupe, 12. Maerz 1885

Lieber Neffe!
Es ist jetzt ein Jahr her, daß ich Deine Karte erhalten habe, und da stand drauf, daß Du bald einen langen Schreibebrief schicken wolltest. Aber wir haben vergebens auf einen gewartet. Ich denke, es wäre bald Zeit, daß Du einen schreiben thätest. Es sind vor Weihnachten 3 Jahr gewesen, daß ich einen Brief an Euch geschickt habe, aber ich habe noch keine Antwort auf denselben.

Wir haben einen außerordentlich kalten Winter gehabt. Ich habe bis jetzt 14 Stück Vieh verloren, und ich glaube, daß noch mehr fallen werden, denn es war kein Gras und die außerordentliche Kälte und Regen mit Glatteis, so daß kein Vieh etwas fressen konnte. Es war nichts zu fressen, wir haben den ganzen Winter das Vieh füttern müssen. Es geht aber nicht mir allein so. Ich bin noch gnädig weggekommen. 12 Meilen von mir hat einer 500 Stück verloren von 1400 Stück, und so geht es fort. Es ist auch kein einziger, welcher nichts verloren hätte.

Friedrich hat voriges Jahr eine Dreschmachine gekauft, 656 Dollar. Er hat gegen 8000 Bushel gedroschen und will auch ein neues Haus diesen Sommer bauen. Karl ist mit seiner Familie auch gesund; er hat zwei Knaben. Auch Christoph ist mit seiner Familie gesund und ich und meine Frau auch.

Der vergangene Winter hat in den Vereinigten Staaten sehr viel Schaden angerichtet, den einen Weg der Schnee und Kälte, und den anderen Erdbeben. Jetzt ist alles besorgt auf Tauwetter. Wenn es stark regnet, werden die großen Flüsse auch wieder viel Schaden machen, denn es hat ungeheuer viel geschneit. Die letzte Ernte war nicht besonders, und deshalb hat das Getreide keinen hohen Preis, der Bushel Korn 50 Cent oder zwei Mark, der Weizen 60–70 Cent, Hafter 30–40 Cent.

Sonst geht es uns gut, ich kann alle Tage Wein trinken. Ich habe letztes Jahr zwei Faß Wein gemacht und hatte noch eins alten, und ein Faß hält 40–

45 Gallonen, die Gallone zu vier Quart. Da ist schon ein guter Tropfen drin. Jetzt wird es aber Früjar, und wir gehen wieder tapfer an die Arbeit. Der Hafer ist schon recht grün. Ich habe auch einen Brunnen machen lassen, denn das Wasser von der Guadelupe herauf holen wollte uns nicht mehr recht gefallen. Er kostet mich zwar viel Geld, aber wir sind doch froh, daß wir einen haben, denn es mußte alles geschoßen werden. Er kommt mich nahe 200 Dollar.
Jetzt will ich sehen, ob einer von Euch Beiden wird Zeit haben und mir auch ein paar Zeilen wird schreiben.
Grüßend Euer Onkel und Tante
C.F. Bergmann und Frau

Ich hatte mir vorgenommen, Euch den Marsch so ziehmlich zu blasen, aber ich dachte, es hilft doch nischt.

Hast Du Gelegenheit, so kannst Du es meinen Schwestern sagen, daß wir noch alle gesund sind.

[16.]
[Handgeschriebener Nachruf für Christian Friedrich Bergmann von dem Germania Farmer Verein; 2. Dezember 1888]

C.F. Bergmann

starb am 24. des Monats gegen 8 Uhr Abends, infolge eines höchst schmerzhaften Magenleidens, an welchem er seit den letzten zwei Monaten krank darniederlag.

Immer weniger werden ihrer der alten Pioniere, die, zumeist in der Hoffnung, ihren Kindern ein menschenwürdigeres Los zu bereiten, als es drüben ihnen zugefallen sein würde, der deutschen Heimat Lebewohl sagten, sich in hiesiger Gegend, welche damals noch wüst und öde lag und ein Tummelplatz raub- und mordlustiger Indianer war, niederließen und dieselbe nach unbeschreiblichen Mühsalen und Entbehrungen in blühende Heimstätten umschufen.

Dem Verstorbenen war es beschieden, seine Anstrengungen, bei welchen ihn seine ihn überlebende Gattin auf das Emsigste unterstützte, in hohem Maße belohnt zu sehen.

Treu und bieder im Umgange, rechtschaffen in Handel und Wandel, furcht- und rücksichtslos das, was für ihn das Recht galt, immer und gegen Jedermann verfechtend, allweg deutsch in Gedanken wie in Worten und

Thaten, dabei gewissenhaft seine Pflichten als Bürger der Union erfüllend, hat er es auch verstanden, gleiche Gesinnungen seinen drei Söhnen einzuflößen, die mit Gattinnen und Enkelkindern einen zahlreichen Familienkreis bilden, in dessen Mitte er patriarchengleich waltete.

Bergmann wurde zu Ebersbach, Königreich Sachsen, im Jahre 1817 geboren. Nach Texas kam er im Jahre 1854, nahm zuerst seinen Aufenthalt in San Antonio, siedelte aber bald nach dem Guadelupeflusse über, wo er sich das behagliche Heim gründete, auf welchem er nun gestorben und beerdigt ist. Das wüste Wetter vereitelte leider die beabsichtigte großartige Leichenfeier.

Beileidsbeschlüsse

Da unser werther Mitbürger, C.F. Bergmann, am 24. November d. J. gestorben, so sei es hiermit von dem Germania-Farmer-Verein beschlossen: Daβ wir in dem Dahingeschiedenen ein werthes Mitglied, einen treuen Freund und einen offenen, biederen und ehrenwerthen Charakter verloren haben, dessen Verlust wir tief beklagen, dem wir ein treues Andenken bewahren werden, und sprechen wir hiermit seiner lieben Gattin und Hinterbliebenen unser tiefgefühltes Beileid aus.
Ferner sei beschlossen: Daβ diese Beschlüsse in das Protokoll des Germania-Farmer-Vereins eingetragen und eine Abschrift derselben der trauernden Witwe zugestellt und in der *Freien Presse für Texas*, *Neubraunfelser Zeitung* und im *West-Texas-Pioner* veröffentlicht werden.

Anhalt, den 2. Dezember 1888
F. Haffring
Hans Wehe
C.H. Hofring
Committee

Bibliography

Benjamin, Gilbert Giddings. *The Germans in Texas. A study in immigration*. Austin, TX: Jenkins Publishing Company, 1974.

Benscheidt, Anja and Alfred Kube: *Bremerhaven und Umgebung 1827–1927. Geschichte im Morgenstern-Museum, Band 1*. Bremerhaven: Nordwestdeutsche Verlagsgesellschaft, 1993.

Biesele, Rudolph Leopold. *The History of the German Settlements in Texas, 1831–1861*. Austin, TX: Eakon Press, 1987; originally published 1930.

Breitenkamp, Edward C. "Seele, Friedrich Hermann." *Handbook of Texas Online*. http://www.tshaonline.org/handbook/online/articles/fse06.

Campbell, Randolph B. *Gone to Texas. A History of the Lone Star State*. New York and Oxford: Oxford University Press, 2003.

Cape, Ruth I. *Youth at War. Feldpost Letters of a German Boy to his Parents, 1943–1945*. New York, NY: Peter Lang, 2010.

Clinton Baker, De Witt. *A Texas Scrap-Book: Made up of the History, Biography and Miscellany of Texas and Its People*. New York, Chicago, New Orleans: A.S. Barnes and Company, 1875.

DeKunder, David, "Anhalt Hall." *Handbook of Texas Online*. http://www.tshaonline.org/handbook/online/articles/xda04.

Eichhoff, Jürgen. "The German Language in America." *America and the Germans. An Assessment of a Three-Hundred-Year History*. Frank Trommler and Joseph McVeigh (eds.). Philadelphia, PA: University of Pennsylvania Press, 1985: 223–250.

Fehrenbach, T.R. "San Antonio, Texas." *Handbook of Texas Online*, http://www.tshaonline.-org/handbook/online/articles/hds02.

Gard, Wayne. "Cattle Rustling." *Handbook of Texas Online*. http://www.tshaonline.org/handbook/online/articles/jbc01.

Genealogical Society of Kendall County. *Keys to the Past. A Journal Compiled & Published by the Genealogical Society of Kendall County* , vol. XXIV/2 Fall/Winter 2005; vol XXV/1 Spring/Summer 2006; vol. XXV/2 Fall/Winter 2006; vol. XXVI/1 Spring Summer 2007; vol XXVIII/2 Spring/Summer 2009vol. XXVIII/1; Spring/Summer 2009.

Gish, Theodore ed. *The Diary of Hermann Seele and Seele's Sketches from Texas*. German-Austin, TX: German-Texan Heritage Society, 1995.

Hander Geue, Ethel, *New Homes in a New Land. German Immigration to Texas 1847–1861*. Baltimore, MD: Clearfield Company, 2002; originally 1970.

Hawgood, John Arkas. *The Tragedy of German-America*. New York, NY: Arno Press and The New York Times, 1970.

Helbich, Wolfgang and Walter D. Kamphoefner (eds.). *Germans in the Civil War: The Letters They Wrote Home*. Chapel Hill, NC: The University of North Carolina Press, 2006. Originally published in German with the title *Deutsche im Amerikanischen Bürgerkrieg. Briefe von Front und Farm, 1861–1865*. Paderborn: Ferdinand Schöningh, 2002.

———. "*Alle Menschen sind dort gleich...*" *Die deutsche Amerika-Auswanderung im 19. und 20. Jahrhundert*. Düsseldorf: Schwann, 1988.

——— ed. "*Amerika ist ein freies Land ...*" *Auswanderer schreiben nach Deutschland*. Darmstadt: Luchterhand, 1985.

Hewitt W. Phil and Terry G. Jordan et alii. *People in Texas*. Boston, MA: American Press, 1981.

Hoecherl-Alden, Gisela and Maedelon Köhler-Busch. "Expanding on Authenticity: A Case for Immigrant Letters, Photographs, and Other German-Americana in the Language Classroom." *Unterrichtspraxis/ Teaching German*, vol. 31/1 (1998): 38–46.

Huebener, Theodore. *The Germans in America*. Philadalphia and New York: Chilton Company, 1962.

Humburg, Martin. "Deutsche Feldpostbriefe im Zweiten Weltkrieg." *Andere Helme—Andere Menschen? Heimaterfahrung und Frontalltag im Zweiten Weltkrieg. Ein internationaler Vergleich*, Detlef Vogel and Wolfram Wette (eds.). Essen: Klartext, 1995.

Jordan, Terry G. *German Seed in Texas Soil. Immigrant Farmers in Nineteenth-Century Texas*. Austin, TX: University of Texas Press, 1966.

———. "*Germans*." *Handbook of Texas Online*. http://www.tsha-online.org/handbook/online/articles/png02/.

———. "*Hill Country*." *Handbook of Texas Online*. http://www.tshaonline.-org/handbook/online/articles/ryh02.

Kamphoefner, Walter D. and Wolfgang Helbich, Ulrike Sommer (eds.). *News from the Land of Freedom. German Immigrants Write Home*. Ithaca, NY: Cornell University Press, 1991; in German: *Briefe aus Amerika. Deutsche Auswanderer schreiben aus der Neuen Welt 1830–1930.* München: Beck 1988.

Kearney, James, "European Immigrant History in the Nineteenth Century." *Discovering Texas History*, Bruce Glasrud, Light Townsend Cummins, and Cary Wintz (eds.). Norman, OK: University of Oklahoma Press, 2014: 94-106.

Kennedy, John F. *A Nation of Immigrants* (New York, NY: Harper & Row, 1964).

Kiel, Frank Wilson. *Civil War Soldiers in Kendall County–A Biographical Dictionary*. Comfort, TX: Skyline Ranch Press, 2013.

Lamprecht, Gerald. *Feldpost und Kriegserlebnis. Briefe als historisch-biographische Quelle*. Innsbruck: Studien Verlag, 2001.

Malsch, Brownson. "Indianola, TX." *Handbook of Texas Online*. http://www.tshaonline.-org/handbook/online/articles/hvi11.

McCarthy, John A. "The German Language in America: An Open Forum." *America and the Germans. An Assessment of a Three-Hundred-Year History*, vol. 1. Frank Trommler and Joseph McVeigh (eds.). Philadelphia, PA: University of Pennsylvania Press, 1985: 270–279.

Moltmann, Günter. "American-German Return Migration in the Nineteenth and Early Twentieth Centuries." *Emigration & Immigration: The Old World Confronts the New,* George E. Pozzetta (ed.). New York, NY: Garland Publishing, 1991: 470–84.

———. "The Pattern of German Emigration to the United States in the Nineteenth Century." *America and the Germans. An Assessment of a Three-Hundred-Year History*, vol 1. Frank Trommler and Joseph McVeigh (eds.). Philadelphia, PA: University of Pennsylvania Press, 1985: 14–23.

Morgenthaler, Jefferson. *The German Settlement of the Texas Hill Country*. Boerne, TX: Mockingbird Books, 2007.

Reichstein, Andreas V. *German Pioneers on the American Frontier. The Wagners in Texas and Illinois*. Denton, TX: University of North Texas Press, 2000.

Roth, David. "Texas Hurricane History, National Weather Service, Camp Springs, MD." http://www.hpc.ncep.noaa.gov/research/txhur.pdf. 17.

Schelbitzki Pickle, Linda. *Contented Among Strangers: Rural German-Speaking Women and Their Families in the Nineteenth-Century Midwest*. Urbana and Chicago, IL: University of Illinois, 1996.

Shore, Elliot. "Introduction. A New Look at the Nineteenth Century." *The German-American Encounter. Conflict and Cooperation between Two Cultures, 1800–2000*. Frank Trommler and Elliot Shore (eds.). New York and Oxford: Berghahn, 2001.

Spengemann, Friedrich. *Die Schiffe der hannoverschen Weserflotte der ehemaligen hannoverschen, später preußischen Ämter Blumenthal, Lehe mit Hafen Geestemünde, Lesum und Dorum und der früheren adelichen Gerichte Lesum und Schönebeck [von der Frühzeit der Seeschiffahrt bis 1901]*. Bremen: Vegesack, J.F. Rohr, 1936.

Struve, Walter. *Germans and Texans. Commerce, Migration and Culture in the Days of the Lone Star Republic*. Austin, TX: University of Texas Press, 1996.

Szymanksi, Hans. *Die Dampfschiffahrt in Niedersachsen und in den angrenzenden Gebieten von 1817 bis 1867*. Bremen: Europäischer Hochschulverlag, 2011.

Tatlock, Lynne and Matt Erlin eds. *German Culture in Nineteenth-Century America: Reception, Adaptation, Transformation. Studies in German Literature Linguistics and Culture Series*. Rochester, NY: Camden House, 2005.

Trefousse, Hans L., "The German-American Immigrants and the Newly Founded Reich." *America and the Germans. An Assessment of a Three-Hundred-Year History*, vol. 1. Frank Trommler and Joseph McVeigh (eds.). Philadelphia, PA: University of Pennsylvania Press, 1985.

Wurster, Ilse and Charles A. Kettner (eds.). *Die Kettner Briefe: The Kettner Letters: A Firsthand Account of a German Immigrant in the Texas Hill Country 1850–1875*. Wilmington, DE: Comanche Press, 2008.

Index

I

J

K

L

M

N

O

P

R

S

T

U

V

W

Y

Z

New Directions in German-American Studies

Werner Sollors, *General Editor*

It is the purpose of this series to subject the large topic of German-America to new critical scrutiny. It does so as an international collaborative effort among scholars in disciplines ranging from modern languages to political history, from American Studies to anthropology, who present independently conceived publications as part of the larger project. Reimagined as part of multilingual America, the new examinations of the German-American tradition in this series offer not only new approaches to German-American studies, but they also force new thinking about what constitutes "German literature" and what have been the defining, though too little recognized, multilingual features of "American literature."

For additional information about this series or for the submission of manuscripts, please contact:

Prof. Werner Sollors
Dept. of African and African American Studies
Barker Center
Harvard University
12 Quincy Street
Cambridge, MA 02138
sollors@fas.harvard.edu

Or

Acquisitions Department
Peter Lang Publishing
29 Broadway, 18th floor
New York, NY 10006

To order other books in this series, please contact our Customer Service Department:

800-770-LANG (within the U.S.)
(212) 647-7706 (outside the U.S.)
(212) 647-7707 FAX

or browse online by series at:

www.peterlang.com

www.ingramcontent.com/pod-product-compliance
Ingram Content Group UK Ltd.
Pitfield, Milton Keynes, MK11 3LW, UK
UKHW020208290926
15470UKWH00001B/76

9 781433 126529